NIGHT

By Cynthia Wall, KA7ITT

Greetings from Oregon
Cynthia Wall
KA7ITT

**Published by the
American Radio Relay League
Newington, CT USA 06111**

Artwork by Sheila Dianne Somerville

In Memory of my Father
Bob Jensen
W6VGQ

CONTENTS

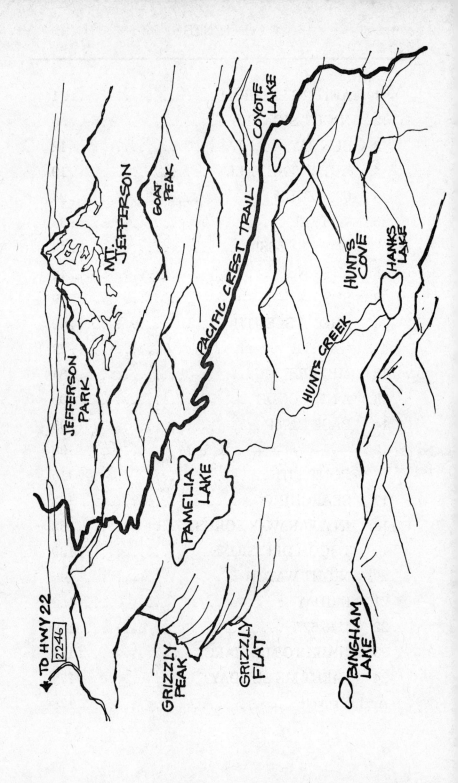

Chapter 1

AFTER THE PROM

Saturday, May 13th
Midnight

It seemed as if the evening would never end. Kim sighed as she leaned back in the white wicker chair near the wall. The ballroom was decorated to look like a southern plantation. Pale green and white crepe paper streamers festooned the sparkling crystal chandelier that transformed the soft lighting into the illusion of drops of starlight dancing across the polished oak floor. Most of the couples were dreamily swaying to the music, holding each other closely as the final hour of the senior prom evaporated into history.

Kim sighed again and swirled the fake mint julep in the plastic glass. What an evening! She had danced perhaps four dances at most with Jeff. All of her attempts at conversation had failed, and now he had excused himself and left the room. She wasn't wearing a watch, but she guessed he had been gone at least half an hour. Where could he be? She felt both angry and embarrassed. Why had he asked her to the prom anyway if this was how he was going to treat her?

Finally, she saw him walking through the open glass doors that led to a veranda and the country club golf course beyond. He casually sauntered over and sat down.

"Can I get you something more to drink?" he asked. He asked it as calmly as if nothing were wrong—as if it were natural to go off and leave your date at a prom.

"I was wondering where you were," Kim said.

"Oh, I just decided to go for a walk—it was getting kind of stuffy and hot in here," Jeff said.

Kim asked quietly, "Have I done something wrong, Jeff? I mean we've hardly danced and then you walk away and leave me here."

He seemed surprised at her question.

"Oh no, Kim, you're just fine. I guess I'm just not big on dances. To be honest with you, I never planned to go to the prom, but my mom told me I had to—said I would be missing out on a great experience if I didn't. I hope you're having a good time, though."

He said it so innocently, Kim almost laughed. There was no answer—no way to retrieve the evening from being a disaster.

"Oh yeah," she said, "I'm having a wonderful time."

He didn't seem to notice her sarcasm—in fact, he seemed oblivious to her. To Kim's surprise, Jeff asked her to dance the last dance, and when the music finally ended, he seemed in a good mood. She looked around the room—her friend Andrea was standing motionless with Tim's arms around her—her head nestled against his shoulder. No one made a move to leave the dance floor except Jeff.

"Well, guess it's time to go," he said cheerfully. "You want a pizza or anything?"

"No," Kim said softly. "I think I'd better get home."

They drove home in silence other than a couple of comments about their plans for summer jobs and college in the fall. Jeff pulled his 1970 Mustang in front of Kim's house. She waited a moment, feeling that she wouldn't be surprised if he didn't even bother to get out of the car.

But he did. He came around and opened the door for her like a gentleman. I wonder if his mother told him to do that too, Kim thought angrily. They walked to the door without saying a word to each other. Kim turned the key in the front door quietly and opened the door partway. Turning to Jeff, she said, "Well, I guess it's goodnight then." He nodded and then leaned forward and gave her a quick peck on the cheek. She said "Bye" and closed the door behind her.

What an unromantic ending to an unromantic evening, Kim thought as she leaned against the cool white birch paneling in the hallway for a minute. She tiptoed to the kitchen and opened the refrigerator. The dim light inside the refrigerator cast a pale glow on her ice-pink formal, making the silky material look even more lovely. "Like cotton candy," she had said when she and her mother had picked out the dress two weeks ago. She unpinned the carnation corsage which was already beginning to look wilted and placed it carefully back in its box on the top refrigerator shelf right beside the bologna and mustard.

"That's about how exciting my date was too," she muttered, closing the door. "About as dull as a bologna and mustard sandwich. Well, it's all your fault," she lectured herself silently as she walked softly to her bedroom. "You wanted to go to the senior prom and when no one asked you, you decided that going with anyone would be better than not going at all."

But, it hadn't been—all those couples intertwined, dancing in the crystal ballroom at the local country club. And, there she had been with Jeff—with them both holding each other stiffly at arms' length, unable to find anything of mutual interest to talk about. Kim wondered why she had ever said "Yes" when Jeff called her. She really knew him only by sight—that tall dark-haired boy who sat two seats ahead of her in American Government. But, she had said yes and then had spent the next two weeks fantasizing about what the romantic prom night would bring. Now she knew. Boredom. She had tried countless subjects—asked him about himself and his hobbies. All of his answers had been in monosyllables. He didn't even like to dance. His only interest had been in the refreshment table and outside walks!

"Kim, honey, are you home?"

Kim winced as her mother turned on the light in the bedroom.

"Yeah, I'm home, Mom." Kim said, trying to sound normal even though she could feel tears close to the surface.

"Well, how was it?" her mother said sitting down on the edge of the bed, obviously ready for a late night chat.

"Oh, okay, Mom. Look, I'm really tired; could we talk about it in the morning?"

They had been too close for too many years for Kim to fool her mom.

"Was it that bad, Honey?" she asked sympathetically.

"Oh Mom," Kim sobbed, sitting down next to her mother and gratefully accepting the warm comforting arm around her shoulders. "It was all just so nothing—and I looked forward to this dance so much. I tried to make him have a good time—I really did—but it seemed like I wasn't even there to him."

They talked for a half hour or so, and when Kim's mother finally said goodnight, Kim carefully took off her pink dress and hung it in the plastic store bag. She put on her cotton pajamas and padded silently to the bathroom to brush her teeth.

"You're a real beauty," she told her tear-swollen face sarcastically in the mirror. She took the hairpins and ribbons from her hair and let her shoulder-length curls flow naturally.

"You know, Kim," she said to her reflection, "actually, you aren't that bad. This isn't the end of the world—there will be other dances and other dates...but never another senior prom," she said, her voice breaking into tiny sobs.

Kim turned out the bathroom light and tiptoed back to her bedroom. She walked to the window by her bed and looked up at the star-studded sky. She sank down on the bed, still staring out the window.

"Somewhere out there in that huge world," she told herself, "there's someone who will feel the same way I do about things —someone who will care for me and I for him—somewhere...someone."

Chapter 2

NIGHT SIGNALS

Sunday, May 14th
2:00 AM PDST
0900 Zulu

Kim was just about to close her eyes and try to sleep when the soft luminescent dial on her International Time clock caught her eye. It was on the shelf directly above her Amateur Radio equipment. She got up and tiptoed across her room. She felt so alone tonight—the thought of reaching out to converse with a stranger suddenly appealed to her.

She thought about how she had planned on telling Jeff all about her hobby of Amateur Radio—about how she talked to people all over the world by voice and Morse code. She had hoped he would be amused at her call letters "KA7SJP." Her ham friends had immediately dubbed her "Sweet Junior Petite" because she was only 14 and five foot two at the time she got her General class license. She had grown only an inch since then, and the nickname had stayed with her.

She had envisioned Jeff coming over to the house, sitting by her side, and proudly introducing him to people on the air. She could have shown him all the QSL cards she had on her walls—postcards that confirmed conversations she had had with other stations. She was proudest of the one indicating she had been one of the lucky stations to hear the Space Shuttle when it flew overhead with Amateur Radio operator Owen Garriott on board.

Kim also had a card from the radio room aboard the ship *Queen Mary* in Long Beach, as well as many from around the world. But, she and Jeff had never gotten their conversation beyond the weather—she doubted he would have been interested in any of her hobbies. He certainly hadn't wanted to share anything about himself with her.

She sat down at the "ham" (nickname for Amateur Radio) rig, deep in thought. There had been a time a couple of years ago when she got up almost every night after her parents were asleep and turned on the rig. With headphones, she could listen quietly to other stations. She turned the monitor volume on her Morse code key down so low that her parents couldn't hear her transmitting. Kim had always preferred Morse code to voice—it was like talking in a secret language, and she was amazed how many people in the world shared that language. Many were the nights that she sat and had conversations with people in Japan, the Philippines, and other Pacific spots far away.

Recently though, the study demands of her senior year had kept her so busy with homework that she scarcely had time to dust her radio equipment, let alone use it. But tonight, the thought of a code conversation with a stranger seemed like a perfect solace to her unhappiness.

She leaned forward and turned on the Kenwood TS-430S, a transceiver (transmitter and receiver), which had been a gift from her Uncle Steve, also a ham, now living in Japan. He was the one who had gotten her interested in ham radio in the first place. He had suggested it as something to occupy her time three years ago when she had had a bad case of mononucleosis and had spent a month home from school.

To her surprise, learning the code and theory had seemed like a game to her and had relieved the boredom of the long days spent at home resting. Her parents had never quite understood her interest in this strange electronic hobby, but they had cheerfully given her the additional equipment she requested at birthdays and Christmas even though her mother had voiced her opinion that she couldn't understand why a girl would be interested in "that kind of stuff."

Kim had discovered that it certainly wasn't an all-male hobby, though. She had talked to many young women hams around the world—some of them close to her own age. And whenever she did have technical problems, she found that other hams were eager teachers. When Uncle Steve last visited, he had shown her how to tune up her 40-meter dipole as a random wire antenna on 80 meters. Two years ago, terms like dipole were Greek to her, but through some dedicated study, she now knew the basics she needed to talk to the world. Uncle Steve's description of the function of a "random" or any length wire as an antenna kind of escaped her now, but she remembered that the antenna tuner served in this system much like a transmission in a car serves between the engine and the drive wheels. The antenna tuner could be made to "match" between the transmitter output terminals and the unknown load presented by the random length wire.

But it wasn't technical information that Kim was interested in thinking about tonight. She needed to talk to someone. She flipped the "on" switch. The TS-430 came alive immediately, the front panel display glowed softly, and a slight reflection of the light from the equipment bathed the ceiling in cheerful shadows. The MFJ 941D antenna tuner performed easily, and soon Kim had achieved a good match to her 40-meter dipole, now connected as a wire working against the water faucet outside her bedroom window.

She put on the headphones so as not to awaken the rest of the house. She was ready to go on 80 meters, the band where she had made her first contact as an Amateur Radio operator.

After she was all tuned up, she debated calling CQ—the internationally known signal that hams use to let other hams know they want to talk to someone. She sat motionless with her hand poised over the key. Instead, she tuned the receiver up and down the band, listening for anyone who might be calling.

Suddenly, —•—•,— —•— There it was—a CQ call. She listened intently to the clear signal which repeated the CQ call five times followed by identification call letters. "This is KA7ITR" she copied on the pad of yellow paper in front of her. The person was sending Morse code at an even ten words a minute, Kim guessed. Whoever it was, he or she was a good operator. The rhythm was clear and steady.

Kim hesitated for just a moment and then switched her rig to transmit. She had been hoping for something more exotic than a "7" call which indicated someone from Alaska, Arizona, Idaho, Montana, Nevada, Oregon, Utah, Washington, or Wyoming. Oh well—maybe a good conversation with someone in Fairbanks or Juneau might cheer her up. Darn, it wasn't even Alaska, she noted as she listened to the call again. Alaskan calls had an L preceding the 7. Oh well, might as well answer.

"KA7ITR this is KA7SJP." She thought of adding (Sweet Junior Petite) but instead just said "Handle (name) here is Kim. QTH (location) is Salem, Oregon."

She smiled—the first time she had really felt like smiling all evening as she heard the reply.

"KA7SJP, this is KA7ITR. Good evening Kim, or should I say good morning? Handle here is Marc. I am nineteen years old. Am backpacking in the Cascades for a few days before taking my final exams at Oregon State University. Did you copy?"

Kim's hand flew to the key. Oregon State! That was where she was going to college next fall.

"Good morning to you too, Marc! I guess it is morning. The clock here says 2:00 AM. I just got back from a dance. What are you doing up in the middle of the night?"

"I couldn't sleep, Kim. All these beautiful stars are keeping me awake. Just back from a dance huh? You didn't mention your age—are you in college or high school by any chance?"

"KA7ITR de KA7SJP. Yes, Marc. I am seventeen and a senior in high school. By the way, I am enrolled at Oregon State next fall. What is your major?"

In her excitement, Kim speeded up her sending to about fifteen words a minute. She realized what she had done just as she switched the transmitter back to receive. Marc's signal came back in clear steady tones at the same speed she had just sent. It was like dancing, Kim smiled to herself. He was following her rhythm and matching it exactly.

"KA7SJP de KA7ITR. Very good, Kim! Hope to meet you in person. My major is computer science. How about you?"

"I'm not really sure Marc, but I'm thinking about veterinary medicine. I'm planning on a biology major for undergraduate work. How are things in the mountains?"

"Beautiful Kim—really beautiful. By the way, do you always go on the air after you come home from a dance? And how was the dance?"

"It was okay Marc—just okay. And no, I don't always go on the air after a date, but I guess I just couldn't sleep tonight either. I used to do a lot of late-night hamming but I have been too busy lately."

She stopped to turn up the volume as his signal was growing fainter and she strained to hear it. She missed part of his next transmission, but she gave a sigh of relief as his tone increased in strength again.

"... is the first time I have been on the air all year. I decided to take a break before finals and see who I could talk to from the top of a mountain—although I'm not really to the top yet, but I will be tomorrow. You are my first contact. I hiked until dark and then spent a couple of hours getting set up, eating, and just thinking. hi."

Kim laughed at the "hi" which is a ham's way of indicating laughter in Morse code.

"Yeah," she sent back rapidly. "I know all about late-night thinking." She paused, blushing. Was she telling Marc too much about herself? She had no idea what he was like. How could she have such a strong feeling from a beeping tone that she was talking to someone she would really like in person? But, the feeling was there—there was no denying that.

"I'm beginning to think maybe you didn't have such a great time at the dance—am I right?" Marc sent. "I'm sorry. I've been to a few parties like that myself—makes you wonder if you are from another planet or something—or maybe you are the only earthling and everyone else is an alien."

Kim waited for him to send "hi" again, but he didn't. He was obviously serious with his humor. She felt comforted. He understood exactly how she felt.

"KA7ITR de KA7SJP" (she had almost forgotten to send her call signal for identification every ten minutes as required by the FCC). "You're reading my mind Marc. Maybe that's a telepathic key you're using instead of a telegraphic one."

This time the laughter came back "hi, hi, hi." She imagined him sitting amongst the tall fir trees laughing. I wonder what he looks like, she thought—probably "ugly as a mud fence" as her grandmother used to say. Somehow, what he looked like seemed very unimportant—right now, he seemed to her like the most handsome man on the face of the earth.

"By the way," she sent. "Where are you?" She had done quite a bit of hiking herself and was somewhat familiar with the Cascades.

"I'm in the Mt. Jefferson Wilderness area north of Detroit Lake," Marc sent. "It's really beautiful up here. I haven't been here for a couple of years, and now I'm wondering why I waited so long to come back. Nice to know there are still places on earth that man hasn't spoiled."

"I agree, Marc. I'm looking at an Oregon map on my wall right now, and I think I see where you are. Do you still have snow up there? What is your elevation? I've never been to Mt. Jefferson itself, but I've been within about 20 miles of there. That's pretty rugged country, as I remember."

"Rugged, but beautiful, and yes there is a little snow, Kim. Just patches here and there, but it's freezing right now. I can feel the ice on the grass next to me. I imagine I'm between four and five thousand feet. I have only seen one person since I started out from my car this morning."

In the first transmissions of their QSO, Kim and Marc spelled out whole words, almost like strangers who talk to each other rather stiffly at first. But, as they began to discuss all sorts of things over the next hour—their interests, their families, their dreams —their words became more and more abbreviated.

"Cu tmw QSO 0200Z?" Marc sent rapidly. Kim agreed to the suggested conversation tomorrow night at 0200 Zulu time.

Kim glanced at her international time clock. Zulu time was a nickname for international time based on Greenwich Mean Time. On a 24 hour clock, it was seven hours later than

Pacific Daylight Time or the same as 7:00 PM Oregon time. She thought quickly of how she had promised her friend Andrea that she would come over and study with her for finals. Well, she could just be a little late.

"FB (fine business)," she sent rapidly, "Lkg fwd 2 tmw nite (looking forward to tomorrow night)." Suddenly she slowed down to her original speed and unabbreviated wording—like a girl lingering on the front step for a goodnight kiss. Marc slowed down too. "Well yl (young lady) I guess I had better let you get some beauty sleep—not that you need it, hi, but I don't want to keep you up all night. Goodnight Kim. 73 (best wishes)." There was a pause and then he sent "88," which is ham lingo for love and kisses.

"Goodnight Marc. Hope you have a fun day hiking tomorrow, and I'll look forward to talking to you at 0200. KA7ITR this is KA7SJP clear."

"KA7SJP from KA7ITR. Sweet dreams Kim."

Kim sat listening for another ten minutes, but the band was silent except for an occasional static noise. She turned off the transmitter and crawled into bed, but sleep eluded her. She sat up and pulled the pale yellow curtains aside to look out the window. It was a clear night, and the sky was covered with stars—the same stars that were shining on Marc—the same stars that were shining on the entire world. She spent a long time watching a bright one twinkle to the east of her house. It seemed to be a slowly moving one—perhaps it was a satellite.

"Goodnight Marc," she whispered and lay back on the pillow. The prom seemed ages ago as she tossed and turned thinking about her conversation with Marc. — — , •—, •—•, —•—• There—that was Marc in Morse code. She smiled as she tapped his name out on her pillow. Somewhere around 4:00 AM, she must have fallen asleep because the sound of the Sunday paper hitting the front porch awakened her about 7. She looked at the clock, thought about her schedule tonight with Marc, smiled, and fell back to sleep.

Chapter 3

SUNDAY

Sunday, May 14th
7:00 AM PDST

T he sun was still hidden behind the mountains, but its early morning light crept along the edge of the meadow at Hunt's Cove like a brush fire trying to catch hold. Passing clouds and misty fog kept the ground temperature at freezing. The spring grass stood stiffly with its nightly covering of frost.

Marc opened his eyes, peeked out from his warm sleeping bag, and saw a young doe watching him from under the tall canopy of the old growth Douglas firs. He stretched luxuriously in the warm down filled bag and thought about his late night conversation with Kim. Here, he had deliberately hiked in away from civilization to give his mind breathing space before the final three weeks of school, and now he found himself anxious to get back—specifically to meet this young maiden of the airwaves in person.

The sun was winning the battle of cold and darkness, so he sat up and watched the warming rays stream through the tall fir trees, dissipating the ground fog. The glistening white slopes of Mt. Jefferson beckoned him. Well, he wouldn't get *that* far, but he would reach the cascading waterfall off to the right of Mt. Jefferson because it marked the location of Coyote Lake.

Marc was more than three miles above Pamelia Lake, and the underbrush was already thinning. His hike in yesterday had been through such lush thick growth that he felt a momentary hesitation at climbing higher. But, the rushing snow-melt streams, ferns, and budding pink rhododendrons were no match for the exhilaration of the high altitude scenery. It was just a short way over the crest now to Hunt's

Lake. He'd lose sight of Mt. Jefferson until he rounded the curve later that morning and scrambled up across the Pacific Crest Trail.

Coyote Lake was an ambitious hike for a weekend. Tomorrow, the eight mile downhill trek back would probably leave him footsore after nine months of little exercise other than carrying books to and from class and his twice a week weight training class. However, he hoped the magnificent ridgetop view from the Cascades would nourish his psyche with energy that would carry him through finals week.

"Our secret place," his father had called Coyote Lake on the few occasions when they had hiked there before. Certainly, it wasn't secret, as other hikers knew its beauty too; still most approached the lake from the longer but easier route of the Crest Trail. Marc and his father had never seen another person past Hunt's Lake on previous trips together. The official trail ended at Hunt's, and after that, Marc would have to cut across the open hillside on his own. This was the first time he had ever gone alone—foolish, he knew his dad would say—but he didn't feel like company.

All last week, even during classes, he thought guiltily, he had mentally planned this trip—where he would go, what food he would take, and most of all—what Amateur Radio equipment he would carry. Low-power ham radio operations and his love of the outdoors made a perfect blend of hobbies. The small seven-pound HW-9 battery-powered transceiver fit easily into his backpack and was light enough to not be an excessive load on long hikes. He had decided to buy a new "Hotshot" 12 volt battery for a power supply that would allow extended operation for his planned four day trip. He had carefully tried out his home-built antenna tuner on a piece of wire in the backyard to ensure that he could "load up" a makeshift wire antenna that he could throw over a tree branch in the forest. Marc enjoyed QRP CW (low power Morse code) operation. He equated it to the challenge of fishing for a whale using three pound test line! He might not work everyone that he called, but when he did snag some DX (long distance contact), it was really a thrill. He had a friend who had contacted Japan from one of these mountains—who knew what goal the magic of hopping or skipping nighttime signals, known as "skip," could reach.

Actually, he had been surprised to talk to Kim last night. She was just 90 miles away and that late at night, he would have guessed that the skip of his signal would have completely passed over Salem. But one never knew—that was one of the fun mysteries of ham radio.

"Time for breakfast," he said aloud to no one in particular, but the answering caw of a black crow seemed approving. Marc pulled on a heavy sweatshirt and jeans over his cotton long underwear before abandoning the warmth of the sleeping bag. His gray wool socks almost came up to his knees and he pulled his lace-up leather boots out of the bottom of his sleeping bag—a trick he had learned from his father to keep boots from freezing. He quickly slipped his feet into them before the morning chill had time to penetrate.

At least it was dry weather. This time of year, rain or snow showers were always a possibility. In fact, as any Oregonian knew, storms were a possibility any day of the year. But, this had been a drought year and so balmy lately that he hadn't even bothered to check the long range forecast.

Marc made a small campfire from dry branches he had gathered from under the tall firs down by the creek. He didn't need much of a fire—just enough to heat some water for washing his face and for a cup of instant cocoa to have along with his granola. He had thought of bringing the works on this trip: bacon, eggs, stuff for biscuits, and so on, but he had discarded those items in favor of a long six cell flashlight (the only one he could find when he packed) and a lightweight bow and arrow which he hoped to use for target practice and possibly for stringing an antenna. A rock would also work for flinging an antenna over a tree branch, but a bow and arrow were a lot more fun as he had found out the night before when he had joyfully shot his antenna up into a tree. He also had a fishing line with a collapsible pole. If he got lucky, perhaps he would have rainbow trout for dinner. Otherwise, he would settle for the dried meat, fruit, nuts, and granola in his pack.

As an afterthought, he had put in one can of Dinty Moore stew and a Hershey's candy bar. He planned to eat them the last night before he hiked home. Sort of a transitional meal from all natural back to the world of prepared foods and preservatives, Marc thought to himself smiling.

His roommate, Mike, had joked of him as he had watched him packing his backpack. It looked like Santa's toy bag as Marc had shoved more and more items into the bulging nylon pack.

"And where's the kitchen sink?" Mike had laughed.

"In the bottom," Marc had replied grinning before he threw a towel at Mike.

"You're going to need a mule to carry that," Mike had warned. He grabbed it and marched down to the bathroom to weigh it on the scale there.

"Forty-one pounds!" Mike announced triumphantly.

"Piece of cake!" Marc said flexing his biceps.

"Yeah, pound cake," Mike retorted.

He might have been right about that, Marc grunted as he looked at his bulging bag of "essentials" lying on the ground. Oh well, it would get lighter as he ate stuff. Within an hour, he was on his way. With one last check to make sure the fire was completely out, he shouldered his pack and headed up the trail toward Hunt's Lake and the high slopes beyond. Today's hike was much shorter than yesterday's five and a half mile trek, but it would include two hours of scrambling up a crumbly talus slope. About three quarters of the way up the slope, he would cross the Cascade or Pacific Crest Trail (a beautiful, scenic trail that runs the length of the Cascades). Then, it was just a short distance to the ridge that surrounded the lake. Coyote Lake was beautiful—pristine in its freedom from human pollution. It was well worth the hike.

Although he knew he wouldn't be actually able to see the sun until it became visible over the ridge about 11:00 AM, he could already feel its warmth in the air. With each degree of heating, it seemed that more and more forest creatures came to life. The light humming of insects was a melody blending into soft bass rustlings from the forest—perhaps squirrels or rabbits gathering food or maybe even an occasional fox or coyote. The forest floor—especially fallen logs—teemed with life. Both the live trees and the dead ones provided habitat for more species than Marc could imagine. Everything from termites to bears lived in these woods. Actually, Marc had seen only one bear in this area, and it had lumbered away before he had time to snap a photo, but he knew they were around. That was why it was so important to keep foodstuffs wrapped up tightly at night.

Bending down, Marc grabbed a handful of sweet clover grass near a trailside flower bouquet of snowdrops, buttercups, and shooting stars. The clover had the sweet taste of ripe green apples, and he savored its flavor as he walked along. The meadow had been one of the reasons he had hiked all the way to Hunt's Cove last night rather than stopping at Pamelia Lake. When he was in a meadow, he felt most like he was part of nature.

He marveled at his own good humor this morning. Less than 48 hours ago, he had been depressed about the possibility of getting a low grade in German class. Hard as he tried to study, it seemed as if the vocabulary words (especially the verbs) just went in one ear and out the other. Schoolwork was far back in his thoughts this beautiful morning. The outdoors always brightened his spirits, but today there seemed to be something more. It didn't take much thinking to figure out what it was—Kim.

Marc thought back to his own senior year in high school. That was when his social life had bloomed for the first time. He had started dating Lisa, a slim dark-haired girl who seemed to be full of boundless energy and fun. They had decided to go steady that spring and were making some tentative plans to go to the University together. The senior prom had been a wonderful evening—the perfect culmination of high school. He felt a wave of sympathy for Kim who had apparently had such a lousy time at hers. He could understand that feeling though. It had been less than a week after high school graduation when Lisa had told him that (1) She had decided she didn't want to go steady anymore; and

(2) She was not going to go to the University with him. Instead, Lisa had applied to and been accepted at a modeling school back East. She had left almost immediately.

At first, they had corresponded weekly; then, the letters got farther and farther apart. At Christmas, Lisa had written saying she had found someone else—someone who seemed to understand her better. Marc had always thought that he understood her okay, but it was obvious that he didn't. He was beginning to think he didn't understand girls at all. Most of the dates he went on seemed dull to him, and so he had given up dating the last few months. He knew he was reading too much into his code conversation with Kim, but it was hard not to think about it. Every time he thought about her, he felt happy.

With sunglasses on to avoid the bright glare from patches of snow on the ground and whistling some melody that he couldn't even name, Marc crossed the partially frozen creek at the end of the meadow and continued up the trail. Nothing like being among the majesty of huge mountains to put one's own life and its trivial problems in perspective, he mused to himself. He had several photos of "Mt. Jeff" hanging on his wall in the dorm, and it had been those that made him think of coming back for a little R & R. He hiked on at a steady pace, anxious to reach the lake by noon so he could spend the afternoon stringing an antenna and setting up camp. Most backpackers he knew rarely came this far because of the difficulty of the trail (if you could call a talus slope a trail) between Hunt's Cove and Coyote Lake. When he reached the top, he would be able to see Coyote Lake, Mt. Jefferson, and the valley below. And, of course it would be the perfect place to work DX.

"Wish you were here, Kim," he said aloud and then laughed at himself. Talk about a blind date. He had never even heard her voice. He imagined it though, and the thought made him pick up his pace—7:00 PM was just ten hours away.

By 10:00 AM, even though he wasn't in direct sunlight, he was already beginning to perspire. He stopped a moment to take off his dacron filled lightweight jacket covering his flannel shirt and stuff it into his backpack.

"Well, here we go," he said aloud as he began his cautious ascent up the crumbling shale-covered slope. The ground was covered with large and small boulders and he used them for handholds as he scrambled like a monkey up the slippery slope. The reassuring splashing noise of the creek below reminded him of the beauty and coolness of the lake above, and he found himself looking up to see if he could see the ridge trail yet.

Where the footing wasn't too bad, he was able to proceed quickly, but in many areas, he had to brace his feet at right angles to each other to avoid slipping backward. He was glad he was wearing leather gloves as the rock surfaces were sharp in places—besides that, they were cold! The slope seemed steeper than he remembered it. He looked down at a clump of large trees near the creek. *I think I was directly in line with those the last time we did this*, he thought. He noted he was a good twenty feet to the left of them. There were no marks on the slope to indicate a trail, and he had forgotten about the trees until just now.

Marc paused—he considered trying to move over the twenty feet, but he was really having trouble keeping his footing where he was. Cautiously, he started up and over at an angle which should move him back on the original path.

He turned to look back down the slope for bearings and felt his heavy pack shift slightly. Then it happened—suddenly his right foot shot out from under him and he found himself sliding feet first down the rock-covered surface. Instinctively, he sat down in an effort to slow his descent, but the crumbly gravel underneath him was like a footing of ball bearings. He skidded for several feet, building up momentum before his left foot jammed into a crevice between two small boulders, and the rest of his body hurtled over the top of the dew-slickened rocks. A loud cracking noise so loud that at first he thought a tree had fallen, echoed in his ears as his body wrenched to a stop with the same force as if he had been in a rear-end auto collision. It felt as if his left knee had practically been driven through the rock. With a scream, he rolled over onto his back, feeling his leg pulled even tighter by the demonic rock snare that had captured it.

Chapter 4

AFTER THE FALL

Sunday, May 14th
Noon

Marc thought he would faint from the pain. A hot poker of agony seared through his leg, and he screamed again and then yelled "Help!" several times as loud as he could. The hollow echo of his panic-stricken voice bounced off the mountain canyon walls. Thoughts, coupled with the urgent messages of pain the nerves in his leg were sending to his brain, raced through his mind. Above all, he was surprised. One moment, he had been climbing upward, and now he was sprawled on his back with his head pointed down the slope. In the many climbs he had taken with his father, he had never done more than twist an ankle. He felt angry at his own clumsiness, but anger was nothing compared to the pain flooding his body.

Sharp rocks dug into his flesh the entire length of his backside, but he was only vaguely aware of that added discomfort. His leg over-rode all other pain sensations.

Because his feet were higher than his head, he had a full view of the weird contorted position of his left leg. Gasping through clenched teeth to stand the surging pain that now seemed in control of his entire body, he raised his head a few inches to scan his injury. He didn't need to be a doctor to know that it was broken—the noise of the bone snapping had startled him even in the shock of the fall. And, he knew that it was badly broken. His entire lower leg was at an odd forty-five degree angle from where it should have been. His foot, sole turned sideways, was jammed in under the rock. He reached out and touched his knee and bit into his lip to keep from screaming again. He tried to roll back on his side to lessen the tension on his leg, but the effort of moving caused him to slip a few more inches down the slope and pulled his tortured leg into an even more excruciating position. Panic-stricken at the thought of skidding further, he ground both hands and his right heel into the slippery rock surface. Every time he raised his head even a little, he could feel the ground slip under him. He had a terrifying vision of himself plummeting headfirst down the hillside like some wild ride at Disneyland.

"Help!" he yelled several more times. He felt a hot sting of tears in his eyes as he heard the fear in his own voice echoing through the valley. He had not seen a soul on the way up other than one backpacker leading a donkey on the way to Pamelia Lake. That man had told him that he was going directly to the Pacific Crest Trail from the lake and heading North. His only hope was that maybe there was a fisherman or hiker somewhere within earshot—perhaps someone on the ridge trail above him. He yelled again and again. Every breath Marc took made the pain crash through his body in huge pounding waves. It was only when he lay perfectly still that he felt able to cope with its intensity.

He was beginning to shiver. Usually a very calm person, he felt frightened and helpless. He was having trouble concentrating. He tried to remember some of his Red Cross first aid training. He wondered if he was in "shock" which he knew was as often a killer of victims as their actual injuries. Marc shuddered with the pain as he raised himself on one elbow and tried desperately to think.

There was a new pain now. His foot and ankle were beginning to swell, and he could feel the laces of his boot cutting into the agonizingly tender flesh. I've got to undo my boot, he thought, but first I've got to free my foot. A wave of nausea and chilliness swept over him, and he dropped his head back on the ground to rest. The pain was quieting a little—as long as he didn't move. He felt dizzy. Passing out would take away his pain, and he felt a strong desire to yield to the temptation.

With a start he forced himself to raise up on his elbow again. "I'm dying," one part of his brain told him, and surprisingly, the thought didn't seem too bad to him—"just let go." But, some other part of him answered angrily "No! Make a plan—do something!"

Whenever he had a difficult problem to solve in school Marc had developed the habit of talking through it aloud step by step. Without realizing what he was doing, he started giving himself instructions in a calm voice.

"Now just relax. First thing is to get that foot free and not fall down the hill any farther when you do it."

Marc suddenly realized that he was still wearing the backpack. The straps had torn on one side, and it had slipped completely over to his right. He could feel the tip of his fishing rod poking into his ribs. Fishing rod. Have to move the rock. His mind seemed to be forming thoughts so slowly. Digging his right foot and left hand deeper into the gravel, he slowly reached back and unzipped the right side of the pack. The collapsible blue fishing rod was on top in the side pocket, and he pulled out the lower end of it.

"This certainly isn't what I thought I'd be fishing for, but let's see if it will work. Okay, now easy does it—just shove the big end of the rod against the rock and PUSH!" He grunted with the effort of pushing the rock. The end of the pole slipped off the rock and made Marc lurch forward. Then his body slipped back, stopped only by the agonizing anchor of his foot. He sucked in his breath sharply and tried again. This time, he made sure the butt end of the pole was directly against the center of the rock. He pushed slowly and steadily so that if the pole slipped, he wouldn't whiplash forward again. Inch by inch, the rock tipped up enough that he was able to drag his foot clear. Moving his leg made him sick with pain, and he turned his head to the side, sure that he was going to vomit. Gradually, the pain eased, and Marc thankfully noted that he hadn't slipped any farther down the slope.

He started talking to himself again.

"Okay, good job. Now let's get that boot off. Let's put the tip of the rod through the knot. Good thing I don't tie double knots. Oops. Okay. Try again. Just wiggle the tip through the knot. Great—it's loosening. Now let's pull on it. Bingo! Okay, now let's loosen the top lace."

He worked carefully for several minutes, jamming the flexible pole in between the leather laces until they were loosened all the way down to the tongue. His face was drenched with sweat from the exertion, and he had slipped a few more inches down the hill, but his ballooning foot was free of the cutting leather. Taking his boot off seemed like an impossibility at this angle, and he decided that it might provide some support for the fractured leg.

"Good job, Marc old boy. Now just call in the rescue helicopter and we'll be out of here in no time." He didn't have the energy to laugh at his own sick humor or his use of the conversational "we" that people often teased him about when he was hamming.

His leg and foot were swelling at an alarming rate. When he fell, it had been his knee that had taken the direct blow that had broken the tibia, the long bone, below it. His knee was bloody and swollen too, and Marc wondered if his kneecap were broken too as he peered down at it through his torn pants.

He ripped the rest of his pant leg open with his hands to ease the pressure on his leg. At least there was no visible blood below his knee, so he guessed the fracture was contained internally.

"I've got to splint that somehow," he said softly, but he had no idea how. He tried to remember the lesson on splints from Red Cross, but his mind was too overloaded with pain and fear sensations to think. He was now shaking and shivering uncontrollably. I really am in shock, he thought miserably. Another flashback from the first aid class. Keep the victim warm—keep the victim warm!

He thought of his jacket and his sleeping bag in his backpack, now lying beside him. Perhaps, he could get them out and cover up. He reached over to pull the pack on top of him and tried to sit up slightly so he could pull his coat out. The movement put increased pressure on his throbbing leg and also caused him to start slipping down the slope again.

With his leg bumping tortuously over the small rocks, he skidded slowly headfirst, still clutching his pack to his chest, a few feet until his shoulders ran into a couple of larger rocks, and he managed to grab hold of them. With his upper body stopped, his legs continued to slide sidewise until they too bumped into the rocks, forcing another scream from Marc. He was now lying horizontally on the slope—still on his back with his injured left leg turned at a forty-five degree angle outwards. He clenched his teeth and lay perfectly still until the pain lessened to a dull roar in his body.

Almost casually, he noted that the sleeves of his flannel shirt were torn and shredded in places, and that his arms were badly scraped. A trickle of blood ran down one forearm and disappeared into the cuff of his ripped glove. He could feel jutting rocks poking into the back of his skull and neck and knew that his entire body had probably been cut and bruised in his unorthodox retreat down the hill. *I wonder why I can't feel that pain*, he thought absently, but dimly realized that the torment from his leg was blocking all other lesser pains.

He was so woozy and felt just on the verge of fainting.

He looked up and saw that the sun was overhead. A red-tailed hawk was slowly circling above the trees scanning the ground for prey.

With a tremendous effort, he pulled his jacket out of his pack and laid it across his chest. With his head and shoulders braced against the rocks, he lay back a moment to think.

It's noon, he thought. *I should be at Coyote Lake.* Unconsciousness was beckoning him again. He vainly tried to resist, but finally he closed his eyes and gratefully accepted the darkness that took away his pain.

Chapter 5

0200 ZULU

Sunday, May 14th
All day

"Sunny today, but a cool Pacific front should bring unsettled weather to the Willamette Valley by tonight. Rain in the lower elevations and possible snow showers in the mountains to 3500 feet tonight and tomorrow."

Kim groaned and reached over to turn off her clock radio. Uh oh, it was already 10 o'clock. She glanced out the window and saw that the car was gone. Her family had left for church. She stretched and settled back onto her pillow. She was glad that they had let her sleep, although how they had managed to keep eleven year old Brandon from waking her was a miracle.

Snow to 3500 feet huh! The end of the much publicized drought. She thought of Marc waking up Monday to a sleeping bag covered with snow and laughed aloud. Oh well, that would give him one day to enjoy the white stuff before he hiked back out Tuesday morning. He would probably have fun tossing snowballs at deer along the trail. She dismissed the thought that a storm might present any problems for him.

She got up and made her favorite breakfast of toast and peanut butter. Daisy, their gray overweight cat, curled up cozily in Kim's lap as Kim sat in the sunny kitchen alcove, reading the Sunday paper and eating toast. She didn't know when she had felt so happy, and the smile on her face was obvious to her parents as they walked in the door after church.

"What are you grinning about?" her father teased, giving her hair a friendly tug.

24

"Kim's in love, Kim's in love," Brandon sang mockingly. Kim just sat smiling at them all. Her mother looked puzzled, and Kim laughed.

She probably thinks I'm crazy, she thought to herself. One moment sobbing in her arms about a lousy prom date and the next moment grinning like the happiest person in the world. Well, maybe I am crazy, but it sure feels good, she told herself.

She felt so happy she didn't even try to chase her brother out of the kitchen as he continued his teasing antics.

"Jeff and Kim...Jeff and Kim...kissing in the car when the lights are dim!"

He danced around the kitchen with his own arms wrapped around his torso like a couple embracing. It was a trick he had learned from his classmates, and it usually brought an angry response from Kim.

Instead, Kim simply smiled and said, "Not Jeff." Her parents stared at her in amazement as she loaded her dishes in the dishwasher and gave Brandon a friendly tap on the head on her way out the door.

Kim spent the morning cleaning her room and listening to tapes. She opened the windows wide and let the sweet smell of rhododendrons fill the room. What a beautiful spring day! She could see clouds gathering in the West though and knew that the sunshine would be short lived. As the weatherman had predicted, it would probably rain by night in the valley. A typical Oregon May day.

She gave careful attention to tidying up her ham station area. She didn't have much room for her equipment, but her father had devised a small computer table in one corner of her room to hold her transceiver, power supply, and antenna tuner. There was enough desk space for her electronic key (capable of sending at fast speeds) and for her old one with which she had originally learned Morse code. An entire wall was covered with QSL cards from stations she had worked, and she had many more in the drawer. She was only two states shy of getting a WAS (Worked All States) certificate. A log book which contained records of her contacts, a *Callbook* which listed names and addresses of hams, and several pads of paper and pencils completed her "ham gear." On an impulse, Kim looked up KA7ITR in the *Callbook* and saw that Marc was really Marc Lawrence of Portland, Oregon and that he held an Advanced class license. Kim just had a General class license, but she hoped to upgrade someday. Every higher class required more theory, and some required faster code skills. Upgrading to Advanced was something she was looking forward to this summer. She glanced out the window at her antenna wire running between two tall black locust trees in the backyard and smiled.

Kim glanced at the bulletin board which held some of her disaster certification cards. Two years ago, she had joined the local ARES (Amateur Radio Emergency Service) group which in her town was affiliated with the Red Cross. She had taken training and received certification in CPR, first aid, and disaster training. The purpose of the group was to provide emergency communication in case of a disaster. She hadn't had much time to do more than participate in the training exercises such as various Walkathons for which the group provided communications.

Oh dear, she noted looking at the expiration dates on her various certifications. She would have to update them all this summer. Suddenly, the thought of working on various "ham" activities with Marc flashed into her mind, and she felt even happier.

After lunch, she went outside to help her mother pull weeds in the early vegetable garden. The two of them knelt in the warm soil and cultivated the rich soil between the rows of carrots and green onions.

When she was sure that Kim's father and brother were out of earshot, her mother turned to her and said, "Okay, young lady, do you want to tell me what all of this is about?"

Kim tried feigning innocence, but she couldn't. She laughed and told her mother all about the late night QSO with her mysterious Marc.

"I don't know how to explain it Mom. I know it's silly to feel this way because I've never even met him, but somehow

I just know that he's really special. Do you think I'm being stupid?"

Her mother put down her trowel and gave her a hug.

"No, I don't. That's the way I felt about your father the first time I met him. I don't know if it's my intuition or what, but there is a certain inexplicable feeling you get when it's the right person. However, I am concerned," she added softly. "I mean, you really don't know this young man, and I would hate to see you get hurt. Try not to expect too much until you actually meet him. Remember how much you looked forward to the prom and how that turned out."

"If I live to be a hundred, it's probably going to always be 'remember the prom,' but you're right—I'll try not to expect too much," Kim promised, but she was already looking at her watch urging it to move forward to 7:00 PM.

At about three, she called Andrea to tell her she would be late for the study session that evening.

Andrea sounded groggy on the phone.

"What did you do—just get up?" Kim questioned.

"About noon," Andrea admitted, yawning into the phone. "Tim and I went over to a friend's house after the prom, and I didn't get home until two. So how was your date anyway? Was he as good as you hoped?"

Kim just laughed into the phone. "Good isn't the word for him Andrea—he was classic," and she giggled again.

"Well tell me about it," Andrea insisted.

Kim paused. The date seemed like ancient history, and she had no desire to make fun of Jeff. It really wasn't his fault anyway. He was who he was, and the romantic image Kim had conjured up of what she had hoped he would be was pure imagination.

"I think I'm in love," she whispered into the phone.

"With Jeff?" Andrea sounded truly surprised.

"No, not with Jeff—with Marc," Kim said enjoying the bewilderment in her friend's voice.

Kim kept the mystery going for a few more minutes, but then her desire to tell her best friend about her "code date" overcame her, and she told Andrea of her conversation and their planned meeting on the air for tonight.

Andrea seemed a little dubious about how someone could feel in love with someone she had just communicated with by beeps, but she listened attentively as Kim related the details she had learned about Marc.

"What do you think he looks like?" Andrea asked.

"I don't know," Kim laughed. "I didn't ask, but I know he's strong enough to climb a mountain!"

"Marc, the mysterious mountain man," Andrea quipped, and Kim laughed with her. They spent a happy hour discussing people at the prom—who wore what, who was going steady, and other items of interest.

Kim promised her that she would be over at her house by eight at the latest and then said good-bye. She thought she heard Brandon hiding behind her hall door—trying to listen to her conversations as always. What a pain a little brother could be! Quietly, she crept to the door and flung it open. Sure enough, there he was looking startled and trying to decide which way to run.

To his total amazement, Kim smiled at him and said "Hi, how would you like to play a game of Monopoly?"

His mouth dropped open in surprise. His sister was certainly acting weird today, but an invitation to play Monopoly, his favorite game, couldn't be overlooked. He came in and for the next hour or so, the two of them fought over control of Boardwalk, the railroads, and so on. Brandon eventually cleaned her out.

"You weren't even trying," he accused her. "Usually you last at least two hours!"

"Nope, scout's honor, you're just too good for me," Kim said.

He looked dubious but gathered up the game and went back to his room to finish painting a model he had started. It was still an hour before dinner, so Kim decided to take Nicki, their three year old German Shepherd for a walk in the hills behind their house.

She put the leash on the excited tail-wagging dog and as an afterthought grabbed her hand-held two-meter rig to carry as she walked. She had bought it used at an Amateur Radio swap meet, but even so it had cost her several months of babysitting money. It was one of her prized possessions, and she carefully attached it to her belt so there was no danger of dropping it if the dog pulled her suddenly. A lightweight microphone fit easily into her left hand, and she grabbed the dog's leash with her right as they set off up the road behind her house.

The two-meter rig was for local contacts only. It sent signals through a mountain-top repeater station (much like telephone companies use) and allowed hams to talk to people in the local area. On a car trip to the East Coast two years ago, Kim had dialed up almost every repeater on the way across the country and had a wonderful time talking to hams along the way.

No one seemed to be on the air now, so she announced her presence by saying "KA7SJP monitoring." Bill, a long time friend and the president of the local club came back immediately.

After a little chit chat about her school plans, she got to the real question she wanted to ask him. Had he ever heard of a KA7ITR? No, he hadn't, but that didn't mean anything because it would be the rare person who knew all of the hams in the area.

He asked her why and she said, "Oh, just wondering. I talked to him the other night and he sounded like an interesting person. Thought you might know him."

She had to sign off then as Nicki spotted a rabbit in the underbrush and pulled her off the trail up an embankment. Kim followed her willingly, and when she got to the top of the hill that bordered a small private lake, she turned the dog loose to run around for awhile. The wildflowers were in full bloom, and their fragrance and color seemed more brilliant than Kim had ever remembered. A cool breeze rustling the trees and the graying clouds scudding in from the West reminded her of the coming storm.

Almost 6 o'clock. She whistled for the dog, and the two of them hurried home for dinner. Her father had barbecued chicken for dinner on the backyard grill. It wasn't warm enough to eat outside yet, so he brought the chicken in, and combined with her mother's corn on the cob, coleslaw, and hot biscuits, the table looked like a summer banquet. The family had always joked about Kim's huge appetite for someone so slender, and so her father automatically served her the largest piece of chicken. Everyone noticed that she left most of it untouched.

"You aren't sick are you, Hon?" her father asked.

"No, I'm fine," Kim reassured him. "The food's delicious, but I just don't seem to be hungry." Her mother winked at her, and her father, intercepting the glance between them, raised his eyebrows questioningly.

"Kim not eating?" Brandon said in a teasing voice. "She must be dying. Quick someone call an ambulance!"

Kim was about to punch him when she noticed that the clock on the dining room wall read 6:45, and she hastily excused herself.

"Just leave the dishes, Mom. I promise to do them as soon as I'm done on the air," Kim said rushing to her bedroom. Brandon, curious about her hurried exit, followed her to her room. She was proud of his progress in learning the code, and usually she would have welcomed him at her side—but not tonight. She considered trying to shoo him away but decided just to ignore him instead. He came in quietly and sat down on her bed as she turned on the TS-430.

She wanted to have time to tune her transceiver perfectly. Working someone on 80 meters at such a short distance was tricky at best, and she certainly didn't want to miss Marc.

"Well, I think that's the best I can do," she said as she switched the rig to receive. She scanned the band but hovered around the place where she had heard Marc the night before. Tonight, there was mainly static. Not a hopeful sign. She waited for several minutes, and then tried calling him.

"KA7ITR de KA7SJP" She repeated her transmission several times and then listened hopefully. Once, she thought she heard a faint code signal, and she eagerly put her ear right up to the speaker. No, it must have been her imagination.

"Who are you trying to talk to?" Brandon asked, but she waved him away impatiently. Finally, he got bored and left.

She kept trying for fifteen minutes—first listening and then transmitting. She could hear some other stations from other parts of the country calling CQ. Normally, she would have jumped at the chance to talk to them, but tonight Marc's was the only signal she wanted to hear.

With every passing minute on the clock, her spirits sagged. Several times, she reached up to switch off the rig, but each time she drew her hand back and tried once again to contact him.

At 0230 International Time, she decided it was useless and turned the power off. She could feel tears in her eyes as she sat staring at the small gray box that could have linked her with Marc.

This is stupid, she told herself. *It was a miracle I heard him last night in the first place. Maybe I'll hear him tomorrow night. Or maybe since I looked him up in the Callbook, he'll do the same for me and call me on the phone when he gets back to school.*

These were the hopeful things Kim told herself, but a small dark thought kept entering her mind. *Supposing last night's contact hadn't seemed very important to him, and he had just forgotten or worse yet, decided not to bother.*

Kim sighed and gathered up her school books to go over to Andrea's. What should have been the happiest time of her high school years—the end of her senior year—was turning out to be very depressing indeed.

Her mother took one look at her long face as she came into the living room and guessed what had happened.

"You didn't hear him, did you dear?"

"No, Mom, I didn't, but don't worry, conditions probably weren't right. I'm sure I'll talk to him again."

She tried to sound cheerful, but she could tell her mother wasn't fooled. She looked as if she were going to say something sympathetic to her, but instead she just asked what time she would be back from Andrea's.

"I don't know exactly. We have a lot of studying to do. Don't wait up for me, and if I'm going to be later than eleven, I'll call you."

Brandon was lying on the living room floor watching a Star Trek rerun and didn't even look up as she left.

Ah, to be eleven again, Kim thought as she started up her ancient VW. *Those were the years when I had no cares.* The wish to return to childhood was short-lived though as she thought about all the fun she'd had in high school. Now was definitely better even if it did include a little heartbreak.

Andrea was waiting on the porch for her.

"Well?" she said expectantly. "Did you talk to him? What did he say?"

"No, the band conditions weren't right, I guess," Kim said. "At any rate, I didn't hear him."

It seemed like too much of an effort to explain skip and 80 meters to Andrea right now, so she tried to change the subject.

"Sorry, I'm late—guess we had better get busy studying for that English test right now."

Andrea looked at her skeptically but didn't say anything. For the next two hours, they quizzed each other on authors, dates, characters, and themes. Kim who was normally a straight A student in anything having to do with literature found herself fumbling for answers.

Finally, Andrea stopped her.

"I can tell your mind really isn't on this, Kim. Want to tell me what you're really thinking about?"

Kim sighed.

"There really isn't much to tell, Andrea. I didn't hear Marc on the air. It's probably because of atmospheric conditions, but maybe it's because he just didn't come on. Maybe, my mom is right. Maybe I expect too much from people I don't really know."

She felt herself close to tears as she made this confession, and Andrea put a sympathetic arm around her.

"Do you know how many times I had my feelings hurt before I met Tim? I really do know how you feel—and let me tell you something Kim. Anyone who would stand you up isn't worth having!"

"But, I don't think he stood me up," Kim protested and then was silent. After all, how could she know what happened? At any rate, she didn't want to discuss it anymore.

"I was just trying to support you Kim—don't be so touchy."

"I know Andrea and I'm sorry. I really am. I just can't explain how I'm feeling now. Forgive me?" Kim asked with tear-filled eyes.

"I always forgive you, Silly. Look, there's a dance next weekend that Tim and I have been invited to. He has a cousin who's going to be visiting him from out of town, and he was thinking it would be nice if he could fix him up with a date. How about it?"

Kim was silent for a moment. Going to a dance with a stranger seemed like the last thing she wanted to do at the moment, but finally she relented.

"I don't know Andrea...oh sure, I guess so. What have I got to lose at this point?"

"Okay, then it's settled. I haven't met him, but Tim says he's good looking and fun to be with. I'll try to get a photo of him for you."

Kim interrupted her.

"Look Andrea, I just don't know. I feel so confused about everything right now. I'm not sure I'd be very good company. Maybe I shouldn't go."

They talked about it for a few minutes and finally Andrea, with an exasperated sigh said, "Kim—at least wait until tomorrow, okay? Maybe things will look brighter to you by then. I can wait until Wednesday to tell Tim if you're not sure."

Kim nodded, suddenly feeling overcome with fatigue. It had been a long, emotional weekend. She told Andrea she would tell her by Tuesday at the latest and apologized for being so indecisive. She gathered up her books, and Andrea followed her out to the car.

"See you tomorrow," she yelled as Kim backed out of the driveway.

The house was silent when Kim arrived. Only Nicki greeted her as she let herself in through the kitchen door. She noted guiltily that her mother had done the dishes. She put her books on the hall table and tiptoed quietly to her room. She set the clock radio for 6:00 AM and then got ready for bed.

As she had the night before, she pulled back the curtains and looked out at the moonlit night. It seemed like an eternity since the night before when she had her QSO with Marc. Then the world had seemed like one magical electrical ball in which anything was possible.

"Where are you Marc?" she said sadly.

She tried to envision him sleeping under tall trees. She couldn't. Suddenly, she sat bolt upright in bed. Supposing, he had missed their schedule for another reason! Supposing something terrible had happened!

Chapter 6

WATER!

Sunday, May 14th
3:00 PM PDST
1000 Zulu

As the warming sun moved through its arc toward the tallest Douglas firs, the still form of Marc Lawrence lying on the slope began to stir slightly. He had no sense of the passage of time as he fought his way toward consciousness. What seemed to him like just a few minutes had really been three hours. The blackness that had been holding him was now interrupted by urgent messages to his brain: pain and something that was upstaging even that—thirst!

Fortunately for Marc, the split ends of the bone had not moved far enough to penetrate the outer muscles and skin of his leg. Marc had gratefully seen that there was no break in the skin above the fracture in his first anguished appraisal of his leg. He had been fearful that he might have a "compound fracture" which could mean infection and gangrene if not treated immediately.

The fracture was severe enough as it was, and the combination of internal blood loss, pain, and fear had caused him to pass out.

But, as his young and healthy body adjusted to the injury, Marc began to regain consciousness. He was shivering, and for a moment he didn't know where he was. His first thought was that he was home in bed. He instinctively reached out a hand to pull the blankets up over his shoulders. The pain of his movement brought him to full awareness.

"Ohhhhhhhh." He let out a long groan and stared helplessly at the blue sky above him. Thirst. He was so thirsty. He reached out to his orange nylon backpack which lay on the ground beside him. The strap of his canteen was poking out of one of several places where his pack was ripped, and he stretched his right arm out to grab it. He only had to pull it out a few inches to see why the backpack was wet on one side. The snap-top lid on the canteen had popped off in the fall, soaking everything around it. He dragged the small plastic container to him anyway and discovered an inch of water still inside. Greedily, he drank it, but it did little to satisfy the thirst that was raking his body.

He was cold, too. Surprised, he looked at his watch which was still running. Another ad for Timex, he thought ruefully. It was 3:00 PM. Even though it would be almost six hours before total darkness, he knew the temperature would start dropping dramatically now that the sun was no longer visible. His coat. He pulled the soft, down-filled orange dacron jacket from the pack, hearing the sound of broken glass and metal rattling as he did. He couldn't bear to look at the damage to his belongings now. First, he had to get warm.

The effort of sitting up enough to slip into his coat was too much, and he finally settled for laying it over his chest as he lay propped against the rock. Once again, his logical brain came into power as he forced himself calmly to assess his situation and try to make a plan. Just the movement of putting the coat over him caused him to suck air in sharply through his teeth to keep from yelling from the pain. His mouth felt as if it were stuffed with cotton, and he tried running his tongue over his parched lips. Water—he had to have water. The splashing of the stream below was an oasis beckoning him.

Marc tried inching little by little down the hill. His broken leg bumped tortuously over the small sharp rocks, and with every jarring motion, he could feel the loose ends of the bones moving inside his grotesquely swollen limb. He stopped, resting up against a larger rock that provided him the security of not slipping any further for the moment. The leg had to be splinted somehow or else he wouldn't be able to move at all.

As he had been moving down the hill, Marc had grabbed one of the straps on his backpack to pull it along with him.

The backpack was fastened to the lower part of an aluminum frame which held both his pack and his sleeping bag. Now, he eyed the aluminum. Could it be used to make a sled for his leg?

Marc shuddered, trying to keep his pain and fear under control. The prospect of getting his sleeping bag and pack off the frame and converting it to anything useful seemed too big a problem to think about. He was having so much trouble thinking anyway. Trying to clear his thoughts, he shook his head.

Then he spied the smaller roll on top of his sleeping bag—his foam sleeping pad. He almost hadn't brought it especially when his roommate Mike had teased him about being a softie. But, his memories of how hard and cold the ground was in May had convinced him to include it. The pad alone wouldn't do it, though. He had to have a splint. He sat staring at the backpack for several minutes.

"How dumb," he muttered. "My brain really is on the blink. There it is right in front of me."

The small unstrung bow with its quiver of arrows was strapped to the right side of his back. There was no way it could have fit inside the pack. Again, his roommate had teased him.

"What are you going to do out there in the woods, Marc? Shoot Bambi?"

"I'm going hunting all right, Mike," had been his reply, "but not for anything to eat."

Marc had remembered hearing someone tell about stringing an antenna with a bow and arrow. He thought it sounded like fun, and, as he had discovered yesterday it was—efficient too. The small bow was one he'd used for target practice as a kid, and he had made a special trip home to get it two weeks ago.

"That just might do," he said now, eyeing the rigid fiberglass bow. But first, he had to supply some padding.

With a groan at the pain the reach caused him, he grasped the rolled foam rubber pad above his sleeping bag. The pad was seven feet by three and a half. Bracing his back against the rock for support, Marc folded the pad in half lengthwise. He rolled slightly over on his right hip so that he could slip one end of the pad under his left thigh. Then began the excruciating process of slowly lifting his left leg little by little to slip the pad all the way under it. When it was done, the pad extended about six inches beyond his boot.

He reached out and quickly untied the cords which lashed the bow to the frame. Carefully, he placed the bow against the outside of his padded left leg, trying to straighten the twisted limb with his hands the best that he could. He was afraid to move it very much, but at least it no longer lay at quite the odd angle that it had. The bow extended down his leg from a couple of inches above his knee almost down to his ankle. Then he set to work. Marc used his belt for the first tie above his knee and then looked around for other suitable wide bands. He was tempted to use the nylon drawstring cords from his jacket, but he knew they might interfere with the circulation in his leg.

I guess the only thing left is to use my sleeves.

He gingerly laid the warm jacket on the ground beside him and looked at the shredded flannel shirt sleeves hanging loosely on his badly scraped arms. *At least I won't have to get my pocketknife out of my back pocket*, he thought as he easily tore off long strips of the already ripped material. His left arm was left completely bare by the time he had made three more tying strips. He secured two of them below his knee, and he strained forward trying to tie the fourth around his ankle. The position was not too unlike warm-up stretching exercises he did in his weight training class, but try as he might, with his injured leg at its odd angle up the slope from him, he couldn't reach his ankle. Every try fueled the fire in his leg, so finally he lay back gasping against the rock. He would just have to leave the lower end of his makeshift splint free and hope his leg survived the journey down the hill.

Afraid that somehow he might slip and become separated from his pack, he reached out and fastened one of the long waist straps to a buckle on his jacket. It would mean that the pack would probably be bumping into him all the way down, but he didn't see what choice he had.

He stuffed his canteen and other things that had spilled back into the pack. He was about to put his coat back in too but decided it might be better to wear it to provide some protection for his bare arm. Besides, it would be getting colder soon even though now he was sweating from all of his exertions. Putting the coat on took a good five minutes and forced several groans from his lips, as he twisted and turned to get his arms into it.

Slowly, he began his descent. Sitting up, he inched his seat cautiously over the rough surface. The rocks were obsidian-like in their sharpness, and he felt as if he were sitting on a bed of broken glass. He kept his right leg bent at the knee so that his heel and his hands provided leverage and some sort of braking power, as he dragged his foam-padded left leg over the ground.

The rocks caught and tore at the foam padding on his leg and quickly penetrated the backside of his jeans, scraping and tearing his tender flesh.

What had taken him about twenty minutes to climb now took several hours to descend. Like a badly wounded crab, he scooted down the hillside, fearful that at any moment he would lose his grip and go tumbling down to the bottom. Several times, he did slide out of control, and each time he dug his heel and hands into the rough surface, bringing him to a precarious stop. His hands were bleeding freely through what was left of his lightweight leather gloves. He tried pulling the cuffs of his coat down to cover his palms, but they kept sliding back up.

At one point, when his leg bumped a rock and sent shock waves of pain through his body, Marc almost passed out. Nausea and clamminess swept over him, and he breathed deeply in and out forcing himself to concentrate on remaining conscious. He inspected his leg quickly and noticed a small bloody area near his shin. There didn't appear to be any bone sticking out, and he prayed that he hadn't compounded his fracture. He remembered that he had a clean handkerchief in his coat pocket, and he pulled it out and placed it carefully over the bleeding spot. By moving one of the straps on his leg a little, he was able to secure it.

Gradually, he started his backwards crawl down the slope once again. The ground was leveling out now. A few sprigs of meadow grass were popping up between the sharp rocks. Gratefully, Marc clutched at them for support as he kept inching backwards. The seat of his pants was practically gone, and he gasped each time he moved over the harsh surface. He glanced at his watch. Eight o'clock. He had missed something—something important. He shook his head trying to remember. Kim. His schedule with Kim that was supposed to have been at 0200, an hour ago. He was feeling so faint again that every thought was an effort.

Kim. Dah dit dah, —•— K, dit dit •• I, dah dah — — M. He sounded her name aloud in code and kept repeating it in a sort of cadence as he backed toward the tall trees and the splashing creek that was growing louder in his ears.

"K dah dit dah —•— move my hands; i dit dit •• move my hips, m dah dah — — move my leg." He kept up the rhythm covering the flat ground toward the stream. There were patches of snow near the stream, and the cold smooth surface was a relief to the shredded skin on his legs and hands. He put some of the snow in his mouth, but still his body craved a real drink of water. His goal was within sight now, and Marc didn't allow himself the luxury of thinking of his pain. He felt so weak and shaky that forming the letters to Kim's name to say aloud was difficult. They came out as a hoarse whisper, but he kept it up, concentrating on the cool water that would soon be his. His pack snagged on small brush, and more than once he had to scoot himself forward to unsnarl it.

At last, he reached the creek bed. The ground next to the water was slushy, and Marc dragged himself through the slippery soil, feeling its cooling balm on his stinging flesh. The actual creekbed was down a slight incline, and he lowered his hips down the twelve inch drop and then carefully lifted his leg down beside him. His pants sopped up water that lapped out from the edge of the stream over the smooth ice-crested stones. He scooted even closer over the slick surface and then rolled his body to his right side as much as he could. Bracing himself up on his right elbow, he thirstily scooped the water up with his hands into his parched mouth.

It felt so good to drink. Marc drank and drank until he no longer felt thirsty. To be sure, the cold water was chilling him, but the discomfort seemed secondary to quenching his need for water. He removed his gloves and rinsed the blood from his hands and washed his face in the clear, icy water. The water lapped along the bottom of his injured leg, and for a moment the fire in his limb subsided. His backpack had bumped down the slope with him, and he carefully filled his canteen and secured the lid. He shoved the pack back up in the grass to keep its contents from getting any wetter.

By now, it was truly getting dark, and Marc was shivering in the cool dusk. Getting back up the slope from the creek proved to be harder than getting down because he had to completely turn himself around and grind his good leg into the soft mud to provide enough leverage to hoist his hips up the embankment. There was no way to drag his left leg up without bumping it on the embankment, and he gritted his teeth with the terrible pain this maneuver caused.

"ISM," (Industrial Strength Marc) one of his friends had nicknamed him because he was calm in times of crisis and always seemed to make decisions confidently. This name flashed through Marc's mind as he dragged his battered body toward a grove of tall Douglas fir trees fifteen feet away.

He fought with all his might both a rising sense of fear and a swirling dizziness. He couldn't recall a truly anxious moment in his entire life. Now, an almost panic-like state captured him.

I'm going to lose my leg; I'm going to starve; I'm going to die. His mind registered these thoughts and then slowly and deliberately rejected them.

"No, I'm not!" Marc shouted, and the strength of his own voice gave him courage. It sounded good to hear himself talking, and so once again, he began giving himself commands.

"Okay. Let's find the right spot. That looks like a good tree to be under in case it rains. And, oh yes, what about an antenna? Because I am going to get out of here," he said fiercely.

He looked up at the tall growth and picked a tree that had a long limb that was fairly free of small branches that might hamper his antenna stringing efforts.

Backing himself up to the tallest tree, he breathed a sigh of relief and rested a moment. It was totally dark now, and he pulled a long flashlight from his pack—another bulky item, but he hadn't remembered to bring his camping lantern from home when he went to pick up the bow. He set the flashlight up against a small rock so that it cast a beam on his legs and backpack and proceeded to try to ready his camp area.

First he unrolled his sleeping bag and spread it out the best he could next to him. The entire backside of his pants was soaked, and he shivered with the cold. His thoughts were slowing again, and he took a couple of minutes deciding whether or not he should remove his jeans. Then he remembered the extra pair of jeans he had stuffed into his pack. Perhaps they were wet too, but he was too tired to pull them out and look right now.

Even though his pants were ripped from the left knee down, he knew there was no way he would be able to slip out of them. He unzipped them and by raising one hip at a time, he pulled them down to his thighs. His scout knife was in his back pocket, and it was easy to reach once the pants were partway down. He pulled it out and easily sliced the denim material from the knee up to the waist on the left side. After taking off his right boot, he was able to pull his right leg out of the remaining pant leg. Now he was really cold. The chill damp earth penetrated his underwear, sending aching cold through him.

The left boot. It would have to come off too. Marc knew that reaching for it with his hands would be too much stress on his leg. Slowly, with his free right foot, he managed to push down on the boot heel and slide it down. His swelling foot had filled the space allowed by his previous unlacing, and it took

42

him several minutes of concentrated force with his right foot to push the boot off and then kick it free.

He unzipped the sleeping bag and laid it open. Gingerly, he put his right leg in first and pulled the rest of his body onto the soft down-filled bag. His injured leg was a throbbing demon, and Marc cried aloud as he positioned it in the sleeping bag. *I bet I ought to elevate that*, he thought dimly as he reached out to turn the flashlight off and then lay back and pulled the warm sleeping bag up over his shoulders.

He thought he would go to sleep immediately, but the pain and cold kept him awake for what must have been several hours. Through a small opening in the tree branches, he could see the star-studded sky, and he forced himself to try and identify the constellations he could see. Clouds were coming in, and soon he only caught an occasional glimpse of the heavens above the gray cloud cover. The wind was blowing, and he could hear the tree branches swaying above him. He shivered, but gradually his body warmth filled the sleeping bag, and he slept.

Chapter 7

POWERLESS!

Monday, May 15th
8:00 AM PDST
1500 Zulu

T he cool Pacific weather front that had been hovering off the coast for two days moved across the valley, dropping light rain on the lush green farmlands. As the clouds moved across the higher elevations of the Cascades, the rain turned to snow. By daylight, temperatures had dropped to well below freezing, and swirling snow blew through the high mountain valleys. The white flakes blanketed the meadow and turned the bumpy ground which had caused Marc so much torment into an ermine carpet.

Under the protection of the trees, only small flurries of snow blew in on top of him as he slept, and he was left fairly dry. A couple of hours after dawn, the snow stopped and a misty ground fog moved in to further engulf the area in quiet whiteness.

Marc woke up to thirst and pain once again. This time, he awoke abruptly and stared in surprise at the eerie shroud all around him. If it hadn't been for the heavy painful reality of his leg, he felt as if he might have been able to float on the mist out through the green borders of snow-trimmed trees and into the valley beyond. He lay perfectly still, awed by the beauty and the stillness all around him.

His four pound down sleeping bag had kept him fairly warm. In fact, his body felt clammy and almost hot from the steamy wetness of his clothes. He put one hand out to grab a little snow and then shivered as a sudden chill struck him. The leg—it was throbbing with a steady intensity. Marc lay still and felt his pulse hammering away in the swollen limb.

He remembered his Judo instructor, Ken, who had frequently talked about the mind's control over the body.

"You are in control; you are the master of your own body," Ken had said.

He closed his eyes tightly and willfully shoved the pain out of his consciousness enough so that he could concentrate. He felt as if he were holding a hungry wolf at bay, and he knew that soon the pain would be back in control.

He looked at his wristwatch. Ten o'clock. He shook his head in disbelief. It seemed like his entire sense of time was disoriented. He wasn't even sure what day it was. He stopped and forced himself to think. Monday. It was Monday. He had casually told Mike that he should be back by Tuesday noon. When would they start to worry? Tomorrow? Two days from now? And what about his folks? Oh, yeah—they weren't home. Had he told anyone at school where he was going? Not specifically. There had been one guy he talked to on two meters in his car on the way up, but he hadn't really told him where he was going. Or had he? It seemed hard to remember. He thought that Kim was the only one he had come even close to giving a general idea of where he was. And, what would she do about it? She probably chalked him up as just another bad date when he didn't show up last night.

Still, even as he thought that depressing idea, a small part of him refused to believe it. Well, even if it were so, he would get on the air one way or the other and simply signal for help.

He lifted the cover of the bag and peered down toward his swollen leg. He couldn't really see it very well and rather than completely uncover in the chilled air, he took the flashlight and aimed its beam at his leg.

What he saw alarmed him greatly. The flesh was grotesquely discolored and bulging from the knee on down, and with the leg lying at its odd angle, it looked as if it didn't really even belong to him. It did though. Any small movement sent pain messages that proved that. He aimed the flashlight at his toes and tried to wiggle them. He could, but it was excruciating to do so. He removed the handkerchief over the

45

cut under his knee. A scab had formed, and he breathed a small sigh of relief. That bloody spot had probably just been due to an external cut rather than a protruding bone. He wondered what more he could do to help his leg.

Last summer, he had worked as a lifeguard at a local pool, and he'd had to take a first aid class to be certified. He tried to remember what the book and the instructor had said about fractures. Most of what came to mind had to do with splinting the limb and then transporting the person to a hospital. What happened if that weren't possible? There's no 911 to call out here, he thought grimly. Any help he got would have to come from himself.

"If it's swollen, elevate it." He didn't know where that quote came from, but it seemed to be a fact in his mind. How could he do that? He looked around him. He was probably going to need everything in his pack right with him. The frame itself didn't promise much comfort. No, he needed something soft. He turned his head and looked behind him under the trees. Several fallen branches lay on the ground within grasp. Or at least, he thought they were within grasp.

Turning his body to reach just three feet beyond him was a feat that strained every muscle in his body and woke up his leg to its awareness of the damage within. Slowly, he pulled two long fir branches to him. He broke the dry wood into several pieces, trying to do so as gently as possible so that the cushioning small branches of needles would stay attached. He built a small elevation on the ground next to his left leg, and then, he reached down and picked up both sides of the sleeping bag and lifted his leg on top of the pile. It was certainly difficult maneuvering the sleeping bag with his body inside it, but he finally got himself shifted so that he was in line with the elevated leg.

It took moments for the pain of moving the leg to subside, but when it did, Marc thought maybe it felt a little better. At least, he felt more able to concentrate this morning, and maybe that was making dealing with the pain easier.

He also felt hungry. *That's a sign I'm still alive*, he thought. He pulled the bag to him and drank from the canteen.

He reached in and pulled out a small apple and some granola—both of which he ate hungrily.

"Now, let's see what's what," he said to himself. Item by item, he emptied his bag. He was relieved to see that his two-meter rig appeared unharmed, but he knew that he was out of range for any of the repeaters from here. Still, it wouldn't hurt to try. He switched it on and was comforted by the familiar rush of static. But, his attempts to raise either the Mt. Hood or the Snow Peak Valley repeater were met with silence. He tried simplex (limited to line of sight or possibly a mile) on the rare chance that there might be someone in range. Silence. He put the rig down on the bag by him and reached into the bottom of his pack to pull out the final items—his 80-meter rig, headphones, and antenna tuner.

The antenna tuner looked none the worse for its tumble down the hill, and gratefully he set it on the sleeping bag. But, the next item, the headphones (his only way to receive on the HW-9 transceiver) appeared crushed. The plastic was broken on one side and missing on the other. He shook them gently and heard pieces rattling around inside. Marc groaned as he envisioned the earphones bouncing between the rocks and his can of Dinty Moore stew all the way down the hillside.

Well, he might not be able to hear, but at least he could still transmit. Fearfully, he reached into the pack and pulled out the HW-9. The small gray box looked okay, and it was still attached to the large 12 volt battery. When he had set the whole rig up in his dorm room right before he left, he'd been so anxious to leave that he had packed as much of his gear connected together as he could. He quickly attached a piece of antenna wire to the terminal on the HW-9 and reached to turn the switch on.

"No!" he shouted in dismay as he saw that the switch was already on. It must have hit a rock on the way down and flipped on. Frantically, he switched it off and then back on again. No S-meter movement on the front panel. He shook it and tried again. Nothing! His battery was dead!

Chapter 8

MISSING!

Monday, May 15th – Wednesday, May 17th

The halls were crowded with students as Kim rushed to her locker to get books for her first period Spanish class. So much had happened during the weekend that coming back to high school felt like returning home after a year's absence. Obviously, she was alone in her feelings as laughter and chatter filled the halls as students shared their experiences of the weekend.

Mementoes of the prom were everywhere. No one had bothered to take down the posters advertising "Some Enchanted Evening—buy your tickets now!" and many girls wore their wilted corsages to school. One boy, who had bought his tuxedo cummerbund rather than renting it, wore it on top of his t-shirt and jeans. The cummerbund was a wild blend of purple and glittery silver swirls and drew rave reviews from the girls.

Kim tried to smile and appreciate all the good humor that was being expressed by her friends, but her heart wasn't in it. Between classes she saw Jeff, and he was mildly friendly as always. *You would certainly never know that we had a date this weekend*, Kim thought to herself. Oh well, she shrugged. That all seemed so unimportant now. The morning dragged on, and Kim tried to go through the motions of paying attention in each class.

There was an essay test in Government, and she was the last one to be finished. It seemed so hard to concentrate that putting her ideas down on paper was a slow process. When she finally walked up to hand in her paper, everyone else had left for lunch. She hurried down the hall to the cafeteria.

Sure enough, there was Jeff on the far side of the room with a bunch of his friends. He didn't even look up as she came in. She went through the salad bar line but didn't see much that appealed to her. She still wasn't feeling hungry.

Andrea and Tim and some other kids that she considered friends were sitting at a long table near a window, so she went over and joined them. They all greeted her, and Tim, with a grand flourish, even pushed a chair out for her with his foot.

Oh dear, Kim thought. *I wonder if Andrea has told him all of my problems? The last thing I want is pity.*

Andrea eyed the skimpy contents on Kim's tray and looked at her questioningly but didn't say anything. Kim sat down and tried to listen to the conversation going on. Now that the prom was over, the subject had focused on graduation and the all night party that would follow it.

Kim couldn't keep her thoughts on what was being said. In her mind, she kept going over and over her code conversation with Marc. Even now, she felt a tingle of excitement when she remembered that hour long exchange. They had talked about so much and had seemed to be so alike in their thinking. She couldn't believe he would deliberately stand her up. No, it had to be band conditions, unless...the small, nagging worry grew inside her. He should be home tonight. Would it be too bold to call him? Well maybe not, if she told him that she just wanted to check on his safety. Anybody ought to appreciate that. Or maybe she should wait until tomorrow.

She debated these questions, but satisfied that at least she had decided that she would call, she turned her attention back to the group.

Kim was grateful for the conversation that took her mind away from her worry about Marc. The afternoon classes passed fairly quickly, and she hurried on home hoping that maybe he would call.

She placed the phone on the desk in her room and busied herself with her homework. Every few minutes she glanced at the clock, tempted to call down to the university.

"You're being stupid," she told herself aloud. "Over-reacting as Mom would say. You ought to know enough about ham radio by now to know that a missed schedule is no big thing." Somehow, even her own parental tone didn't convince her.

The family ate dinner. Brandon entertained them with his usual antics. Kim smiled at him. She got a kick out of her brother's colorful descriptions even if they weren't always table conversation. He was a real comedian, and he had a true gift for telling jokes.

At 6:45, she excused herself from the table and ran to her room. She didn't really expect Marc to be on the air. After all, he was supposed to be back at school, but suppose it had been so nice in the woods that he had been tempted to stay another night? It sure wouldn't hurt to listen for him.

She tuned up the transmitter carefully and listened on the frequency where she had talked to him before. Nothing. She tried calling him several times. No answer. There was lots of static on the air anyway so hearing anyone would probably be difficult.

She sat there for ten more minutes with her ear pressed hard against the speaker, hoping to hear "KA7SJP de KA7ITR." All she heard was static.

Just then, Brandon appeared in her doorway.

"Want to do some code practice?" he asked wistfully.

Kim was about to tell him no, but she looked at his face and said okay. Her homework was all done, and besides, working with Brandon would take her mind off her troubles.

For an hour they sent code back and forth. Brandon was really getting good and was pretty solid at both sending and receiving five words a minute. She quizzed him on some of the theory questions.

"Looks like you've been studying!" she told him approvingly. "I think you're almost ready to take the Novice test."

He grinned, waiting for her to say more.

"Let me check with the club, but I think there are going to be tests given the first Saturday in June. Want me to sign you up?"

Instead of answering with his voice, he sent "roger" rapidly in Morse code. Kim leaned forward and gave him a big hug. They spent some happy time talking about what it would be like when he got his Novice license. Their Uncle Steve had promised to supply him with a rig of his own if Kim took hers off to college.

"Of course, by that time, you'll have your Technician license, and we can talk back and forth on two meters," she told him approvingly.

"If you have a good enough antenna here at home, and I'm able to put one up in the dorm at school, I might be able to talk to you every day," Kim promised him.

The mention of college brought her worry about Marc back to her thoughts, and she told Brandon that it was time for him to go to bed. Reluctantly, he dragged himself out of the room, leaving her alone with her thoughts.

Kim picked up her phone. She hesitated for a minute and then made up her mind. Quickly she dialed information and asked for the number of Oregon State. She called it, and when the campus operator came on the line, she asked for the phone number of Marc Lawrence.

"Just a minute, I'll connect you to that room," the operator told her.

He's going to think I'm an idiot, Kim thought to herself, but she didn't have much time for such thoughts before a deep male voice answered the phone.

"Hello?"

"Hello, I'm looking for Marc Lawrence. Is he there?"

"No, this is his roommate, Mike."

"Mike, I'm an Amateur Radio operator. My name is Kim Stafford. I talked to Marc Saturday night by Amateur Radio, and I'm wondering if he's returned from his backpacking trip yet?"

"No, Kim. He hasn't. He said he would be back tonight, but I figured he probably decided to take an extra day. He only has one class tomorrow, so he might have decided to stay out longer. By the way, do you know where he went exactly? He never did tell me."

"Somewhere in the Mt. Jefferson Wilderness Area, Mike, but I'm not sure exactly where. I was concerned about him because he made a date to meet me on the air last night and he didn't show up."

"Oh?" Mike sounded concerned.

"Well, that's not unusual. Radio conditions have been kind of peculiar lately, so I don't think we should be worried about that. But, if he doesn't come back tomorrow, well then..." She paused trying to think of what to say next. "Look, I want to give you my phone number, and if you hear anything from him, will you give me a call?"

"Sure, Kim."

They said good-bye, and Kim hung up feeling more worried than ever. She went downstairs where her mother was sewing in the family room. She told her what had happened.

"I just know something is wrong, Mom, but I don't know what to do about it."

Her mother sat quietly thinking.

"Honey, I think you just need to wait until tomorrow and see if he comes back. Maybe by this time tomorrow he'll be back and you'll have talked to him on the phone, and everything will be okay."

"Mom, if he doesn't come back by tomorrow afternoon, can I go down there? I mean, maybe I need to talk to the police or something."

"Are you sure you couldn't do that by phone?"

"I'm not sure they'd believe me. I think I need to go in person. Please understand, Mom."

"I do Kim. In fact, I'll go with you if you want."

Tuesday at school dragged by for Kim. She rushed home the minute the final bell rang. Her mother wasn't home from her part-time job yet, and Kim let herself into the house. The light on the automatic phone answerer was flashing, and anxiously, Kim rewound the tape to playback any messages.

"Hello, Kim? This is Mike down at Oregon State. It's 2:00 PM here, and Marc hasn't come back yet. We're starting to get a little worried around here, and I was wondering if you had heard anything? You might give me a call when you get home."

Kim listened to the rest of the messages, but they were all for her parents. She quickly dialed the college, but no one answered in Marc and Mike's room.

Just then, she heard the familiar sound of her mother's car coming in the driveway. Her mother came bustling into the house, her arms laden with paperwork she brought home every night from work.

"Did you hear anything?" One look at her daughter's face told her that she hadn't.

They sat down and talked. Kim was anxious to drive to the school immediately but finally agreed that since Marc was technically only a few hours overdue, that they would wait until morning. It was an anxious night for Kim, and Mike's call woke her at 5:00 AM saying that Marc had never shown up.

"I'm coming right down," Kim told him.

Her mother was already up and said she would go with her.

"If you go to the police, you may need an adult along."

Soon, they were on their way down I-5 to the Corvallis turnoff. The sun was just rising as they left the freeway at Albany and took the back road to Corvallis which wound along beside the Willamette River.

Kim's anxiety was building, and she unclenched her fists and tried to force herself to relax. This is all going to be all right, she said over and over again silently.

She realized her mother had been talking to her—asking her questions about school.

"I'm sorry Mom. What was that? Guess I wasn't listening."

Her mother reached out and patted her arm.

"Nothing Dear. Just trying to take your mind off things, but I guess that's impossible."

They drove through the gates to the campus and stopped at the small information booth to ask directions to Marc's dorm. It was almost 7:00 AM. The campus was quiet. Kim's mother found a parking place at the back of the dorm, and they went into the lobby. The student at the desk rang Marc's room.

In a couple of minutes, the elevator doors opened, and a tall, blond young man walked out. He looked around the lobby, spotted them, and came over with his hand outstretched.

"You must be Kim and Mrs. Stafford. I'm Mike Ringland."

They talked for a minute, and then Mike invited them up to the dorm room. The room was small and cluttered. One of the beds was neatly made, and Kim assumed it was Marc's. She and her mother sat down on the edge of it while Mike talked.

"I don't know what to tell you. Marc had been looking forward to this trip for weeks. He never told me much about where he was going—just somewhere near Mt. Jefferson, he said. I wouldn't really be worried about him except that he's always punctual. The rest of us around here kind of take time commitments fairly casually, but when Marc says he'll be somewhere at a certain time, he's there. I think he's the only guy on this floor to never miss a class."

Kim felt her stomach knot with worry at this latest clue about Marc's personality.

"Do you have a photo of him, Mike? That might be helpful if we go to the police."

"Gee, I'm not sure. Oh wait a minute. Someone took some snaps at the baseball game last week."

He rummaged through a desk drawer and finally produced a small photo which he handed to Kim. It showed Marc standing at bat, his hat on backwards and a big grin on his face.

"That was the only homerun hit all day," Mike said laughing.

Kim stared intently at the photo. Marc looked so much like she had imagined him, that she felt gooseflesh all over. He was tall and had dark wavy brown hair and even white teeth that showed in his smiling but determined-looking face.

"Mike, I know they'll ask us things like height and weight. Do you have any idea," Kim's mother asked.

"Oh, about 6′ 1″ and 170 lbs. He has brown hair as you can see, and I think his eyes are brown too—yeah, they're definitely brown just like mine. I remember one time we were joking about getting colored contact lenses and going in disguise somewhere. I just can't believe this is happening." Mike shook his head sadly.

"Mike, it might help to know what kind of equipment he had with him. Do you remember? And, I mean radio equipment as well as camping gear."

"Yeah, we had quite a bit of discussion about that," Mike laughed. "That guy took so much stuff, I don't know how he could have walked. Sleeping bag, even a pad to sleep on, food—apples, dried stuff, a can of stew, a huge flashlight, even a bow and arrow. Said he was going to string an antenna with it or some crazy idea."

Kim jotted down the items on a piece of paper. Marc was certainly not the average hiker.

"What about radio equipment?" she asked.

"Well, I don't know much about that stuff, but there was something that he had built himself that I know he was pretty proud of—said he was going to talk to the world with that."

Kim nodded. That would be the 80-meter rig—the HW-9 that he had told her about on the air.

"Anything else, Mike?"

"Well a bunch of stuff for his antenna and then a smaller rig I think—one that he uses around here on campus."

"Is it a small gray or black box that he holds in his hand?" Kim asked.

"Yeah, black, I think. I think he has a microphone that attaches to it too, but sometimes he doesn't use it—just holds the rig right up to his face—you know the kind that security cops use."

"I sure do, Mike. That's a two-meter rig, and I'm glad if he has that with him. I don't know if he could reach any repeaters from where he is, but we can certainly pass the word out on a net and start listening."

They were about to leave when Mike said he would like to go to the security office with them.

"Maybe the more of us there are who think there is a problem, the more likely they'll be to start a search," he said.

They walked briskly across the campus to the small office which said Campus Security on the door.

A balding, middle-aged man was sitting at the desk eating a doughnut and drinking a cup of coffee.

"Hi, can I help you?" he greeted them.

Kim quickly explained the problem. Mike joined in a couple of times, adding that he was very concerned.

The man listened intently.

"Have you checked with his home?" he asked.

Kim said no that they hadn't.

"Well, let's do that first." He dialed the registrar's office and got Marc's home phone number.

"No answer," he said after he had let it ring ten times. "Okay, let me tell you what I recommend. Generally, we say to wait 24 hours after you think someone should be somewhere and then contact the County Sheriff. But in this case, I can tell you are really worried, and there's a possibility that he may be injured in the woods somewhere. I think we should phone the sheriff's office from here so if need be, both you and Mike can talk to the dispatcher to give information.

He walked over and studied an Oregon map on the wall.

"I'm not sure exactly what county he's in—looks like it could be one of several, but let's try Marion for starters."

He dialed and after a few brief statements, he handed the phone to Kim. The woman's voice sounded cordial but very business-like.

Kim felt herself almost stammering as she relayed the story.

"Hello, I'm an Amateur Radio operator, and I had a conversation with a university student who was hiking in the Jefferson Wilderness Area. We talked on Saturday night, and he said he would talk to me again on Sunday night and that he would be back at school late Monday. Well, it's Wednesday now, and he didn't show up for our schedule and he hasn't come back to school. His roommate is here with me, and we both think that something has happened to him."

The dispatcher paused a minute. Kim could hear her writing on a pad of paper. She asked Kim for her name, address, and phone number. Then she started asking questions about Marc. What was his age? How tall was he? How much did he weigh? What was his home address and phone number? Kim was able to answer all of the questions except the date of birth and Mike didn't know either. She stopped a minute to confer with the security officer, and he quickly dialed the registrar on another line to get the information.

What was he wearing? What equipment did he have with him? Kim handed the phone to Mike who wasn't sure of all the answers.

"Well, his jacket is orange, I know that, and he was wearing blue jeans. About his equipment, I can't be too sure. I watched him pack some of it, and I know he had a lot of radio gear plus a sleeping bag and oh yeah, a bow and arrow. I'm not sure what else. I came in the room just as he was finishing and I teased him about how much it weighed, but I never did actually look in the pack."

"How much did it weigh?" she asked.

"Forty-one pounds—I know because I took it down the hall to weigh it."

"And do you know how much of that was radio gear?"

"No, I don't—maybe ten pounds I would guess, but I'm not sure. I saw it all one night when he had it lined up on the floor."

He put Kim back on the phone and she related what she knew about his transmitting equipment.

"He told me about it on the air. Hams always do that because they're usually proud of their equipment. It was a battery operated HW-9. I don't know how much that would weigh, but if it's portable, probably not too much. Why is that important, anyway?" Kim asked.

"Well, if we have a general idea of how much his other gear weighed, it may help us to know how well prepared he was for survival."

Kim shivered thinking of him. It was a lot cooler than even yesterday, and she was sorry she hadn't worn a jacket down here.

What about his car, the woman wanted to know. Mike told her that it was a '78 blue Chevy Malibu.

"Kind of a scratched-up paint job—he was always talking about getting it repainted but he never had the money. And yeah, I know the license number. It was his call letters…let's see."

"KA7ITR," Kim supplied.

The dispatcher asked Mike some more questions about Marc. Did he have any health problems? No, Mike didn't think he did. Was he an experienced hiker? Mike said he had mentioned hiking some with his dad, but that to his knowledge he hadn't done any this last year.

It was very important that they get hold of Marc's parents. Did Mike know where they were?

"Well, I did hear him say something about them taking a trip somewhere, but I'm not sure if they're back now or what. I'm afraid we didn't talk much about that sort of stuff."

The dispatcher questioned Kim some more about any clues Marc might have given in their conversation as to his location.

"All I know is that he said that he was at about 3500 feet and that he would be going a lot higher the next day. Oh yeah, and he said there were some patches of snow on the ground."

"Will you be at a phone where you can be reached?" the woman asked Kim. Without even thinking about school, Kim said that yes she would. Mike said that he would be in class, but he left a number where he could be reached and promised to check in with Kim every day.

Kim thanked the dispatcher and then hung up. It was silent in the tiny office, and the four of them sat for several minutes just thinking. The quiet hum of the electric wall clock seemed to permeate the whole room.

"Try not to worry too much," the security officer reassured her. "These things happen all the time, and most of the time, the missing person turns up the next day with a perfectly good

58

reason for being gone. Your friend sounds like a responsible person."

Mike nodded in agreement, but the look of worry didn't leave his face.

They walked with Mike back to the dorm.

"Promise me you'll let me know the minute you hear anything," he said as they reached their car.

"You bet and you do the same for me," Kim said.

They shook hands, and then Kim and her mother got into their car and drove slowly out of the dormitory parking lot. Mike stood on the sidewalk watching them until the small blue car rounded the tree-shaded corner and disappeared from his view.

Chapter 9

SOS ••• ——— •••

Monday, May 15th 11:00 AM PDST
— Tuesday, May 16th 7:00 PM PDST

Marc sat dazed for more than an hour after he discovered that his battery for the HW-9 was dead. This just couldn't be happening to him. His last hope was gone.

It wasn't his nature to accept defeat though, so slowly, forcing himself into some sort of action, he frantically shouted "Help" at the top of his lungs every few minutes for the next hour. It was no use, and the effort exhausted him.

He pulled the sleeping bag up around his shoulders and shivered in the cold morning air. *Eat something*, his brain commanded him. He pulled a handful of Gorp, the popular dried fruit and nut mixture, from the side pouch of his bag and munched on it slowly. Gradually, his shivering eased, and it became easier for him to think.

Without really knowing why, he spread all of the contents of his backpack out in front of him.

"Maybe some brilliant idea will come to me, if I just see what I've got," he said, but his voice lacked conviction.

His roommate had been right. He had brought a lot of stuff, but it certainly didn't seem like the right stuff now. He looked at it all as he piled it on his sleeping bag in front of him and also to both sides of him in the snow. He had to be careful not to put anything on top of his painful left leg.

Finally, it was all spread out, and he did an inventory of sorts.

One quiver of arrows with two bow strings
Two pairs of jeans (one wet and one dry)
Two pairs of underwear and socks and one extra t-shirt
Collapsible fishing rod and extra line

Roll of antenna wire
The HW-9 with its now useless battery
Antenna tuner/SWR meter
2-meter rig
Repeater Directory
Boy Scout pocketknife
Wristwatch with second hand
Toothpaste and toothbrush
Washcloth, towel, soap, toilet paper, and comb
Wallet and car keys
Six cell flashlight
Plastic canteen and set of cooking and eating utensils
Two books of damp matches
Book of German grammar
Three apples, four small bags of dried fruit and nuts
Two packages of dried meat and potatoes
Four squished rolls
Two packages of dried soup
Three packages of instant cocoa
Hershey's candy bar

One can of Dinty Moore stew (which he thought had probably been responsible for breaking things in the fall).

Supplemented by fresh trout, he certainly should have had enough to eat well for his three day venture. But to make it last for who knew how long...Marc shook his head and wondered. He was suddenly ravenously hungry, and he quickly ate one of the rolls and drank a little water from the canteen.

Cold, he was so cold! *I'll build a little fire next to me*, he thought. He gathered the twigs that were within reach and crumpled a couple of pages from the *Repeater Directory* (directory of two-meter repeater stations). The match covers felt damp though, and Marc fervently wished he had thought to seal them in a plastic bag or to bring waterproof ones. All of his radio gear preparation had been so well thought out, but his camping gear had been a little haphazard. He tried to strike several of the matches. One flickered briefly but not long enough to light the paper. Rather than waste them all in a futile attempt, he put them back in the pack. If the sun came out, perhaps he could lay them out to dry. Then remembering a trick of his father's, he pulled one pack out and put it on top of his head underneath his wool cap. He was glad he had brought the wool cap, as he had slept in it last night.

Nothing seemed like a possible solution to Marc as he lay contemplating his possessions. If the clouds ever cleared, and if he heard a plane, perhaps he could signal with his flashlight. Or maybe a hiker would come this way. In the later summer months, there were hardy souls who used this area as access to the Pacific Crest Trail. If only one of them would come now. *If, if, if.* He beat his fist angrily in the snow beside him.

He was sleepy again. Was he just going to slip into a coma and die? He fought the warm feeling slipping over him, but finally he closed his eyes and dozed. When he awoke, it was late afternoon.

He propped himself up and set to work filling his canteen as it was almost empty. He tied about twenty feet of fishing line onto the strap around the neck of the plastic jug and then tied a small rock to the strap to weigh it down once it hit the water. It was slow going. His first two tosses didn't come even close to the stream.

"And I'm the pitcher of the dorm baseball team," Marc muttered, discouraged.

The effort involved in getting a really good throw put a tremendous strain on his leg, and he groaned as he hurled the bottle once more. But this time, a small splash greeted his ears, and he could feel the bottle sinking at the end of the line. He waited a few minutes and then gradually began reeling it in, hoping that not too much of the precious contents would spill on the way.

It was about half full when he got it back to him, and after taking a long drink, he capped the bottle and put it back in his pack.

Dinner. Should he try to eat one of the packets cold by mixing some water in it or should he wait? He might need to ration whatever he had. He finally compromised by mixing a packet of cocoa with some water, and he consumed that plus one more roll.

The clouds were clearing as the sun set, and Marc lay awake anxiously awaiting the sound of any airplanes. About two hours after dark, he heard one in the distance, and frantically he began flashing SOS with the flashlight. The sound of the airplane engine grew fainter and fainter until finally Marc could not hear it.

The leg felt as if it were on fire. *It's like a wild beast that wakes up every few hours*, Marc thought. *It hurts all the time, but sometimes it really roars.* It was roaring now, and Marc grimaced as he tried to turn the leg slightly to relieve some of the pressure.

He tried watching the stars for awhile, but somehow the bright lights of the sky seemed as if they were connected with the nerves in his leg. He closed his eyes and to take his mind off the pain, he forced himself mentally to send code. Without realizing it, he found himself having an imaginary conversation with Kim.

Dit dit dit dit H dit e dit dah dit dit l dit dah dah dit p dah dah m dit e. "Help me," his brain whispered. *Dah dit dah K dit dit i dah dah m—*"Kim."

Gradually, the pain eased, or at least he hypnotized himself against it, and he fell asleep.

Sunlight streaming through the trees woke Marc early on Tuesday. The warming light was falling on his face, and he lay still enjoying the heat.

Maybe I won't freeze to death after all, he thought gratefully. No fire, but the sun is an even better source of energy.

"Another source of energy!" He said it aloud with so much force that he sat bolt upright. "That's what I need—another source of energy!"

Quickly, he unloaded his pack again in front of him and began talking to himself.

"Okay, my battery is dead. What else do I have? Only the flashlight batteries, and they're too small." He paused and then excitedly he said, "Or are they? There are six of them.

Each one is one and a half volts—that would be nine volts total—I wonder if that would run the rig?"

Frantically, he unscrewed the metal cap on the bottom of the flashlight and jammed a piece of antenna wire into the metal threading around the edge. Then, he put the end back on and re-screwed it tightly, catching the wire firmly.

"That's my minus connection," he said. "Now, how do I get my plus lead?"

He unscrewed the top of the flashlight and looked at the plus side of the top battery facing him.

"This just might work," he said unscrewing the front lens assembly. He unscrewed the plastic bulb holder and positive contact assembly, allowing the bulb to drop out. Then he took a rock and broke the glass of the flashlight so he could push a wire through the hole so it came in contact with the positive end of the battery. He screwed the top back on to hold it firmly in place.

"I've got both positive and negative—now let's hook them up to the HW-9. I wonder what my high school physics teacher would have thought of this arrangement. I don't care what he thinks—I'll give myself an A plus if it works."

With shaking fingers, he attached the wires plus a short antenna wire. Closing his eyes in fervent hope and prayer, he switched the transmitter switch on. The S-meter wiggled in response to the power, and Marc let out a whoop of joy.

"I did it," he yelled to the listening woods and mountains around him. "I did it!"

His voice was quivering with excitement as he kept on talking to himself.

"Okay, my headphones are smashed, so I can't listen. But, if I tune this thing up correctly, I should be able to tell if I'm getting out. But, how will I know if anyone hears me? When they rescue me, I guess. Wait a minute, Marc. Let's not talk about tuning it up until you've got an antenna."

His shivering, both from cold and excitement, was so intense, that he lay back for a moment and covered himself up to regain a little warmth. Even with the exhilaration of his

64

Cross Section of Marc's Modified Flashlight to Provide Emergency Power Supply to HW-9 Transceiver

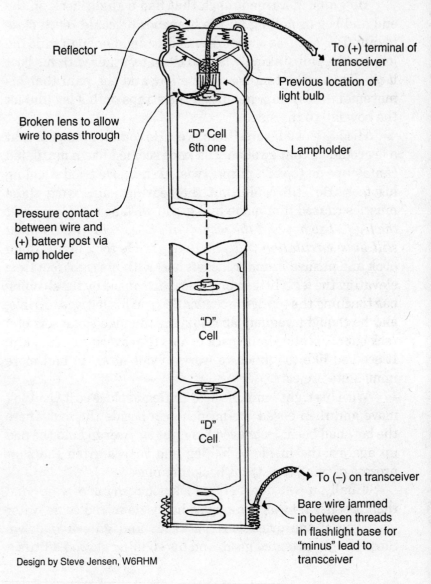

Reflector

To (+) terminal of transceiver

Previous location of light bulb

Broken lens to allow wire to pass through

"D" Cell 6th one

Lampholder

Pressure contact between wire and (+) battery post via lamp holder

"D" Cell

"D" Cell

To (−) on transceiver

Bare wire jammed in between threads in flashlight base for "minus" lead to transceiver

Design by Steve Jensen, W6RHM

invented power supply, Marc could feel sleep beckoning him like an evil ghost. It would just feel so good to close his eyes.

"No!" he shouted and sat up again. He reached into his pack and drew out his quiver of arrows. "Oh no," he groaned looking down at the bow strapped to the side of his leg. Well, this was more important. He would just have to find something else for his leg.

He grabbed a tree branch that had a slight hook on the end and began pulling all the branches he could reach close to him. Two or three of them seemed straight enough to do the job, and he painstakingly lashed them together with his boot laces. Then, sweating from the effort and the pain that his movements cost him, he untied the straps on his leg and let the bow fall to the side.

His leg looked worse than ever. He couldn't imagine that a leg could be that swollen. His knee seemed like a mutilated cantaloupe on top of a purple twisted stalk. He tried wiggling his toes. He still could, but the searing pain using those muscles caused him made him groan with agony. *How could the leg feel so hot and the rest of him be so cold? Well, I must still have circulation in it,* he thought. He took his German book and pushed it under his left heel with his right foot thus elevating the leg a little more. The pain caused by the sleeping bag touching the top of the tender flesh on his leg was terrible, and he thought vaguely about trying to make some sort of a rack out of his aluminum pack to hold the covering off his skin. It seemed like too much to worry about now. He had more immediate projects.

With his right hand, he held the leg in place so it wouldn't move and then eased the branches in beside the pad where the bow had been. He moved his right leg over to hold the pad up against the inside of his leg and then started the slow process of tying the straps back into place.

Finally, it was done. Marc took a long drink of water and then because he was shaking so badly, he mixed some water with one of the dried food packages and gulped it down. Surprisingly, it tasted good, and his shaking slowed a little.

He pulled one of the bow strings from the quiver and strung it tightly on the bow. Done. Next, he tied the end of his lightweight monofilament fishing line to the end of an arrow. He wanted a higher antenna than he'd had on Saturday, so lying back against the tree trunk for support, he took careful aim at a branch about thirty feet above him.

Zing. He let loose the arrow and watched it fall short of its target and plummet back to the ground. He dragged it back to him and tried again. The next time, the arrow snagged on a lower branch that was jutting out from the tree, and Marc wrestled with the line, jiggling it and pulling on it, for several minutes before the arrow fell free.

On the fifth try, the arrow sailed over the desired branch and came to rest dangling about fifteen feet down. Marc raised himself up as much as possible so that he could give a large whipping motion to the fishing line with his right hand. It was slow work, but luck was with him, and the line was not snagged on anything. Gradually, the arrow dropped farther and farther down until finally, Marc could reach up and grab it. He quickly cut the fishing line and tied his antenna wire to it. Then slowly, so as not to risk snagging the wire, he pulled the heavier antenna wire up into the tree.

When the antenna had reached the level of the branch, he tied the fishing line still attached to it to a rock on the ground to secure it as much as possible.

Ground—he needed a ground wire. Once again, his throwing skills came in handy as he hurled a light rock with a ground wire attached into the stream. He attached the other end to the transmitter.

Now he started the process of tuning up the HW-9, trying to get as good a match to his antenna as possible. Each time he made an adjustment, he pressed down on his code key, and when RF output on the meter was midscale, he was satisfied.

"Well, here goes," he said solemnly. He began slow and methodical sending. "SOS ••• S — — — O ••• S. SOS ••• — — — •••"—he sent the International Morse code distress signal over and over. Then, hoping he had attracted someone's attention, he sent his call letters and the message "injured in

Oregon Jefferson Wilderness area near Coyote Lake." *Are you hearing me, Kim?* he wondered. If only there were some way to tell if anyone were hearing him. He glanced at his watch. *6:00 PM or 0100 Zulu (International Time). Would Kim still be listening at 0200 for their supposed schedule of two nights ago? Or was it three nights ago? What day was this anyway?*

Panicked at the thought that his brain wasn't able to function properly, he forced himself to recount all the events of the past 48 hours.

"It's Tuesday—I'm sure it's Tuesday," he said. He pulled out his pocketknife and made a small notch in a piece of bark beside him. Everything was getting hazy again. He wanted to fight that feeling, but on the other hand, sleep was the only thing that took his pain away. And, he was really hurting. He gritted his teeth and looked at his watch. 0200—how could a whole hour have passed? It just seemed like a few seconds.

"I'm going crazy," he whispered and reached out to the key beside him.

"SOS SOS SOS SOS" His hand fell off the key and he dozed. When he awoke with a start, it was almost eight. His transmitter was still on. *Wasting power!* he thought anxiously and switched it off.

Just then, he was caught in a series of convulsive sneezes that shook his whole body. His head felt hot, and he drank the last of his water from his canteen before lying down and pulling the sleeping bag up around him.

Chapter 10

PHONE SLEUTH

Wednesday, May 17th
9:00 AM PDST
0900 Military Time

Sergeant Bill Willis whirled his chair around and studied the topographical maps spread out on a long table beside him. "Someone possibly lost in the Jefferson Wilderness Area" was the report that had just come in this morning. The dispatcher had given the report to the Field Supervisor Shift commander. Now, it had been passed to him as Emergency Services Coordinator.

He sighed and looked at the large expanse of mountains, gullies, lakes, and rivers that covered the four-county wilderness area. It would be like looking for a needle in a haystack if they didn't get more specific information.

He finished off the lukewarm coffee in his mug that said "Have a nice day," and rubbed his eyes. It had been a long three days. On Sunday, a three-year-old child had wandered away from her parents in a picnic area. Panic-stricken, they had called the Sheriff's Office. Fortunately, that had been a simple search, and searchers found the child scared and cold, but unhurt less than a half mile from the picnic area. They never could figure out why she hadn't heard her parents calling unless the noise of the cascading stream had blocked out their voices. It was a miracle that she hadn't fallen in and drowned somewhere.

The next case had been more troublesome. A middle-aged couple had rented a cabin for a weekend and taken along their 75-year-old mother who had Alzheimer's disease. Her condition wasn't so far advanced that she couldn't enjoy an occasional outing, and so they had taken her up to see the woods. They had only left her alone for half an hour while they went for a walk down to the store at the lake to buy some things for dinner. She was napping in the bedroom with their Golden Retriever, "Max," at the foot of her bed. He had sat up eagerly when he heard them open the door to go, but they had firmly told him to stay there and take care of Grandma. She had just gone to sleep, and she always slept soundly for at least two hours in the afternoon, so they felt they could leave her safely.

When they got back, the front door was standing open and both Grandma and Max were gone. They had run into the woods in all directions shouting for Grandma and whistling for Max. Nothing. That was when they had called the Sheriff.

That search had just ended late last night. Searchers had combed the area for two miles in all directions. No one thought she could have wandered farther than that. After the second day, hope grew dim as temperatures dropped. She'd worn only a cotton dress, lightweight sweater, and bedroom slippers at the time of her disappearance. Everyone involved in the search feared that a night of sleeping in the snow would probably kill her.

But, they hadn't counted on Max. They found Grandma in a little cave near a waterfall about three miles from the cabin. She had been curled up on the ground tightly against Max who only stood up and started barking when he heard the rescuers walking through the woods whistling for him. Grandma had a slight case of frostbite on her hands and feet, but other than that, she seemed none the worse for her experience.

"Well, that's two out of three," Sergeant Willis said. He had a funny feeling about this case though. Somehow, he thought it was going to be a difficult one.

The call had come in less than an hour ago. He had already sent an inquiry out on the teletype to local law enforcement agencies in neighboring counties seeking information on both the vehicle and Marc "the person of interest." The preliminary facts didn't sound too alarming—a hiker who was one day overdue, and his due back date might not have been too accurate to begin with. The guy was young and healthy, and the two kids who had called it in had thought he had quite a bit of gear with him.

He looked over at the display board leaning against the wall. "Ten Items for Survival" it read—a visual aid for speeches their department gave to hiking clubs. Pasted on the board were such items as firesticks, a whistle, tea bags, maps, compass, mirror, and rope.

I wonder how many of these Marc Lawrence has with him? Sergeant Willis thought as he looked at the missing person's form that the dispatcher had filled out.

He studied the weather reports that had just been laid on his desk. The snow of the beginning of the week had stopped and temperatures had actually risen to above freezing in the mountains yesterday. They had dropped to about 25 early in the morning, and another storm system was due this afternoon that could bring more snow. Snowfall in the Cascades had been practically nil the last two months, but it looked as if that situation were rapidly changing.

Time to get busy. He picked up the phone and called Kim. Her mother answered and said that Kim (protesting) had gone to school after they'd returned from Corvallis. It was her mother's day off, and she had promised to stay home by the phone. No, there hadn't been any word from Marc, and Mike, the roommate, hadn't heard anything either. Kim's mother had also tried the parents' house again, and there had been no answer there.

Sergeant Willis thanked her and said he would be checking back during the day.

Time to activate the forces. He had already put various search and rescue agencies on "yellow alert" which meant that there might be another search coming up. Now, it was time to act.

He phoned the Sheriff's volunteer Jeep Patrol to start the search of the road heads to try to locate Marc's car. They were an effective group, and if Marc had left his car anywhere on a passable road, they would find it.

He studied the information he had on Marc again. He was an Amateur Radio operator; in fact, that was how Kim, the girl who had reported him, had come to meet him. So he had radio gear with him. Seems like someone sooner or later would make contact with him that way. Sergeant Willis made a note to himself to talk to Kim later in the day about alerting ham nets to listen for him. The dispatcher said she sounded like a pretty smart girl. Maybe she had already done that.

He picked up the phone and dialed Marc's parents once more. To his surprise, someone answered.

"Hello?"

"Hello, is this Mrs. Lawrence?"

"No, it isn't. This is her neighbor, Marjorie Bradock."

"Mrs. Bradock, this is Sergeant Bill Willis from the Marion County Sheriff's Department. I'm trying to reach Mr. or Mrs. Lawrence."

There was a brief pause and then a very worried voice said, "Oh my, is something wrong?"

"Well, we're not sure, Mrs. Bradock. Are the Lawrences at home?"

"No sir, they're not. They're out of town. I'm feeding their cat and watering the house plants for them while they're gone."

"When will they be back, Mrs. Bradock?"

"In about a week, I think. What's wrong?"

"Well, Mrs. Bradock, their son Marc is overdue back from a hiking trip. Some friends of his reported him missing, and I'm just trying to gather some more information."

"Oh my. That's terrible, just terrible. Oh, his parents will be so upset."

"Is there any way we can reach them Mrs. Bradock?"

"I don't think so. You see they went to Europe for a month, and they didn't leave any numbers for me to reach them."

Sergeant Willis tried not to groan aloud. This was certainly a dead end trail.

"Mrs. Bradock, do you know of anyone who might know their whereabouts? Any other relatives, friends?"

The line was silent a moment as she thought.

"June, Mrs. Lawrence that is, plays bridge with a group of friends. I know a couple of them by name. Do you want me to call them?"

"Well, actually, if you have their phone numbers, I would appreciate it if you just let me phone them. I have very specific information I want to ask."

Mrs. Bradock looked up the phone numbers of two of the women in the group and gave them to him. The sergeant asked for her own number so he could check back with her.

Margaret Sims, the first woman, didn't answer her phone, but Joyce Maynard, the second one, did. He quickly explained the situation.

"Oh no, how awful!" Mrs. Maynard exclaimed.

"Well, we don't know whether the boy is in any trouble or not at this point, Mrs. Maynard, but we're just trying to take precautions. Do you have any idea of the Lawrence's itinerary in Europe?"

"She did talk about it some at bridge club. All three of them were so excited about the trip."

"All three of them?" Sergeant Willis interrupted.

"Yes, Mrs. Lawrence's mother went with them. They had several friends in Europe they wanted to see so that was why they weren't going to take a tour. I'm afraid June didn't give me any specifics though—just mentioned that they would be driving through France, Austria, and Germany."

"Do you have any idea if they made their reservations through a local travel agent?"

"Yes, that I do know. In fact, one day while we were having lunch together downtown, June stopped off to pay for the tickets. It's ABC Travel on Center Street"

"Thank you very much Mrs. Maynard. That information may be a big help."

Sergeant Willis got up and refilled his coffee cup. He could tell this was going to be a long morning. Thank goodness, he had been able to go home for a few hours of sleep the night before. Even so, he felt tired.

"ABC Travel, may I help you?"

"This is Sergeant Bill Willis from the Marion County Sheriff's Department. We're involved in a search for the son of one of your clients who is currently vacationing in Europe. We're trying to locate them and hope you can give us some help."

"What are their names?"

"Bradley and June Lawrence."

"Just a minute, please."

Sergeant Willis sighed as the all too familiar "Muzak" came on the line while he waited. Shortly, an agent, who identified herself as the one who had sold the tickets, came on the line. He explained the situation to her.

"I'm afraid I'm not going to be too much help to you. They left here on May 1st and flew to Frankfurt. They're due back on the 31st. The only hotel reservation I made for them was in Vienna and that was six nights ago."

Sergeant Willis sighed again as he looked at the calendar. May 17th—Marc's parents weren't going to be back soon enough to be of any help.

"How about a car rental?" he asked.

"No, they were going to take the train part of the way, and they also were going to be staying with friends, I believe. They said they would rent a car if they needed it."

Sergeant Willis thanked her and hung up. He called the Lawrences' neighbor, Mrs. Bradock. Did she know of any relatives the Lawrences had?

"Well, Mrs. Lawrences' mother went with them. Her father is dead and so are both of Mr. Lawrence's parents. I believe Mr. Lawrence has a brother somewhere back East, but I'm not sure where."

"Mrs. Bradock, the next time you go over to their house to feed the cat, will you just glance around to see if maybe they left an itinerary somewhere? I'm not asking you to search through drawers or anything—just see if there might be something lying on top of a desk."

"Yes sir, I will, but I must tell you, I already did that after you called this morning, and I didn't see anything. I'll look again though."

Amazing! Sergeant Willis thought, as he hung up. *How could people go away and not leave numbers where they could be reached? It happened all the time though. And why hadn't this Marc fellow told someone where he was going? Kids!* he thought. Well, he had two teens of his own and he understood being impulsive. He smiled a little as he admitted to himself that he had been the same way at that age—in fact, even with all of his training, occasionally he still didn't take all of the precautions he should.

He called down to the university. Mike was in class, but he reached a resident advisor in the dorm. He seemed to know Marc pretty well and had heard that he hadn't come back as scheduled.

"That surprises me. It really does. He always seemed like the person who was on time to everything. I think I've heard other kids say that he never missed a class."

Did the advisor know anything about Marc's gear that he took or his backpacking ability?

"No, afraid I can't help you there. I didn't see him when he left, and he never talked to me about backpacking."

Sergeant Willis had an idea and called Mrs. Bradock back.

"Sorry to bother you again M'am but we're kind of running out of leads here. How long have you lived next door to the Lawrences?"

"About ten years sir."

"Do you know the names of any of Marc's close friends?"

"Let me see. He was a popular kid—always had a lot of friends over to the house. There was one boy though—Kevin Miles—down the street that he used to have over a lot. I know because I used to hire the two of them to mow my lawn."

"Do you know Kevin's phone number?"

"Well, he went in the Navy, but I can give you his mother's number."

He thanked her and then sat thinking a minute before dialing the Miles' residence. This was a wild goose chase.

"I'll bet Kevin will be out on some remote ice island or something," Sergeant Willis muttered.

Not quite, but close. Mrs. Miles answered the phone and told him that her son was aboard a nuclear powered submarine somewhere in the Arctic.

"I'm not sure what it would take to reach him, but I suppose we could try," she said.

"Well, maybe you can help me, Mrs. Miles. Did your son ever go backpacking with Marc?"

"No, I don't think he did. I believe Marc and his father went on those trips alone. Kevin always had a newspaper route so he was never free to leave early in the morning."

"Do you have any idea where they went on those trips? Did you ever hear them talk about them?"

"Oh, I don't think they went that much. Marc's father traveled a lot, and I think he just went camping with Marc to kind of make up for the time he was gone. I do remember him mentioning a trip to the coast once and I think a couple of times they went somewhere up near Detroit Lake, but other than that, I'm afraid I really don't know."

"Detroit Lake? Thank you Mrs. Miles. That may be quite a bit of help to us."

He hung up and looked at the maps again. Detroit Lake was a favorite camping and fishing area off Highway 22 which traversed the Cascades into Central Oregon. The trails that led from the area were numerous, but at least the lake was a starting place. He went to the radio to give the Jeep Patrol this new information.

On his way back to his desk, he picked up an updated weather report.

"Heavy morning ground fog. Increasing clouds and lower temperatures. Lows of 20 – 25F in the Cascades with accumulations of 5 – 6 inches new snow possible tonight."

Chapter 11

"KA7ITR"

Wednesday, May 17th
9:00 AM PDST

Sharon Hansen, better known to her ham buddies as WB7RHO, had just poured herself a second cup of coffee. She looked out her kitchen window at the gray, foggy morning. Her sweatshirt over her long-sleeved shirt felt good as she waited for the fire she had built in the woodstove to warm the room.

She had lived in Oregon for forty years, but each year it always amazed her how it could be cold and wet so late in the spring when other parts of the country were posting temperatures in the nineties. Well, that was what gave Oregon its reputation for liquid sunshine and people rusting instead of tanning. This year had been warmer and drier than usual, but today looked like "typical Oregon weather." She looked out the window and admired the blooming rhododendrons, azaleas, and peonies which decorated her yard with brilliant color.

She read the morning paper as she sat at the kitchen table waiting for the phone call that she expected. Half an hour ago, Jeri, the secretary of the Jeep Patrol, had called, telling her that they were on "Yellow Alert" for a possible search. A male college student was possibly lost in the Jefferson Wilderness Area.

Whenever there was a possible search, Sharon anticipated it until it started. Her mind and body were as ready for action as was her 4-wheel drive truck parked in the driveway. She always kept the truck packed with emergency gear—warm clothing, food, first aid supplies, maps, water, and most important of all—radio gear. The truck was equipped with a high-band radio with which she could talk to the Sheriff's Dept. and other Search and Rescue Agencies. However, she also carried her two-meter ham rig, extra battery packs, and antennas so that she could be in touch with the other ham members of the Jeep Patrol...and, if necessary, with ARES (Amateur Radio Emergency Service) members who might be providing radio communications for the area.

The shrill ring of the phone startled Sharon even though she had been expecting it.

"We've gotten the go-ahead for the search," Jeri told her on the phone. "Everyone's supposed to meet at the parking lot by the Timbers Restaurant at 10:00."

Sharon's husband had already left for work. She turned off the coffee, put the cat out, grabbed her coat, and within five minutes, she was on her way across Salem heading for Highway 22 which led up to the meeting place near Detroit Lake. In the early morning fog, the few cars on the road all had their headlights on and were driving slowly.

The fog lifted a little as Sharon drove through the foothills, and except for a few dense patches, it was easy driving. She monitored the radio and checked in with a couple of other members of the Jeep Patrol who were also on their way to the rendezvous place. She turned on her two-meter ham rig to see if any of her friends were on frequency. A couple of hours ago, the repeater probably had been humming with hams "ragchewing," commuting to jobs.

Her mission this morning was a lot more urgent than a regular job, and Sharon felt her heart pounding in anticipation of what the day might bring. She always felt a combination of dread and excitement when she was called out to help on a search. The first one she ever went on was ten years ago after a fellow ham suggested that she join the patrol. That one had ended happily—they had found the car of an overdue hunter, and searchers had found him badly injured about five miles from his car. Their timely rescue undoubtedly saved his life, and from then on, Sharon had realized the importance that every member of the search and rescue teams played. She was proud to be a small part of the effort.

This morning though, she had an uneasy feeling. Perhaps, it was because she too had a son about the age of Marc who was also away at college. She had only a cursory description of the lost boy, but in her mind she saw him as David, her own son. She wondered how well the lost student had equipped himself in preparing for the wilderness. Most of the time her own son was very level-headed in his planning for anything, but there were other times when youthful impulsiveness had gotten him into difficult situations. What kind of young man was Marc?

"KA7ITR"—they were looking for a fellow ham. That fact made her feel an even closer bond to this unknown person. Like Sharon and so many other hams, Marc had personalized license plates that proudly bore his call. Oregon, like other states, supplied Amateur Radio operators with the special plates for just a few dollars extra in appreciation for the public service they performed.

The information Sharon had gotten about Marc said that he had been heard on 80 meters. She didn't have 80 meters in her car, but she did at home. She flipped on her two-meter rig again. From past experience, she knew that once she got very far into the hills, she would be out of range of any repeaters. If they got close enough to him, they might hear him on simplex—that is, if he had a two-meter rig with him, and the description the roommate had given of his gear indicated that he probably did.

The forest-bordered pastures were dripping with moisture from the fog, and Sharon watched the grazing cattle and sheep that were slowly moving through the wet grass. She liked this time of day when the world was quiet. The day might prove to be tension filled and chaotic, but this time was hers.

Two of the patrol members were already waiting for her as she pulled off the road into the parking lot by the convenience market across the street from the restaurant. They were standing by the open tailgate of one member's truck, drinking coffee and studying maps of the area. Sharon had her own maps which showed every road and trail in the area. They would divide the area to be searched into one mile square sections. Every road and side road in the area would be searched. Probably just five or six of them would be involved in this preliminary search. If they failed to turn up the car, then other agencies would be called in.

Sam, who was head of their unit, waited until everyone was there, and then they discussed the best place to set up a temporary base camp. Since it was midweek and there weren't too many cars on the road, they decided the parking lot was as good a place to set up as any. One person, their commander, would remain here and monitor the frequency for their reports every half hour. He would be in contact with the Sheriff's Office.

The wilderness area comprised over 100,000 acres, and they had only a few clues to go on. Marc had transmitted that he was somewhere in the Jefferson Wilderness Area and that he was near 4000 feet. With over 160 miles of trails in the area including 36 miles of the Pacific Crest Trail, the search could be huge. The mother of one of Marc's friends remembered something about Marc and his father going to Detroit Lake. That often was a stopover or departure point for hikers, so they doubted if that lead had any real significance. If they could just find his car, the search would be dramatically narrowed.

It was decided that Sharon would work the southern edge of the wilderness perimeter and explore all of the roads leading into it. She would gradually work her way west until she met up with John who would be coming her way.

And so started the search. Sharon stopped at the edge of the first road that led off Highway 22 and got out. She carefully anchored a piece of colored (to contrast with the snow) toilet paper on a tree branch at the left side of the road, indicating that she was in the process of searching. A two inch covering of snow blanketed the ground, and Sharon made sure that the paper was visible on top of some green tree branches poking through the snow. About a mile down the dirt road, she stopped and put her truck into four-wheel drive to avoid slipping off the road. There were several side roads, and she had to explore each one. Her eyes scanned the ground carefully, looking for any clues.

The Jeep Patrol members all assumed that they were looking for his car, a '78 Chevy Malibu, but from experience, Sharon knew to look for anything. It was always possible that the boy had left his car somewhere else and hitched a ride up here. In that case, they would be looking for any clue—footprints, a piece of camping gear, even a candy wrapper—anything that indicated that a human being had passed this way recently.

The snow obliterated any footprints, but at one clearing where the snow had melted in the direct sun, Sharon got out and walked around the edge of what was a parking area in the summer to see if she could spot any marks.

Her hopes rose as she spotted tire tracks on the next road, but then in a flash she realized that since they were on top of the snow, they were too recent to be Marc's. Still, she followed them to the end of the road. Sure enough, a red Toyota pickup was parked there, and two sets of footprints led into the woods. Probably some day hikers. If she could have intercepted them, she could have asked them to be on the lookout for Marc, but they were long gone.

It was clouding up again and the wind was blowing. She felt grateful for the warmth of her truck as she climbed back in. 5.4 miles down this access, and she had driven down three side roads for a total of 11 miles off this road alone. Since morning, she had explored over 100 miles of road. She paused at the junction with Hwy 22 and got out to put an additional piece of toilet paper on the other side of the road, indicating that the road had been searched and nothing found.

No one else was finding anything either. They checked in with Base Camp every thirty minutes. Everyone reported the same. Very few people in the woods this time of year—only two cars found on road heads, and neither one of them was Marc's.

It was mid afternoon when she turned right onto Road 2246 labeled Pamelia Creek. It had been recently paved and the driving was fairly smooth. She worried as she drove along.

There were only a couple of roads left in her area to search, and it would be dark soon. Where was this young man?

Then she saw it. The blue Chevy was sitting at the edge of the road clearing, under some tall fir trees. Snow had covered the top of the car including the windows, but the license plate was clearly visible. KA7ITR.

"I've found the car," Sharon radioed quickly to her commander.

She knew it was her job to secure the area so that no evidence on or near the car could be disturbed. She backed off about 25 yards and watched for Sergeant Willis who was enroute.

While she waited for the deputy, she searched the perimeter of the area, to look for footprints. If he had taken a trail, the one leading to Pamelia Lake was most likely.

Sergeant Willis arrived on the scene and joined her in scouting the area. It was starting to snow lightly, and they brushed the falling flakes from the Chevy's windshield to look inside. The faded upholstery was bare except for a map lying upside down on the front seat. He quickly produced a "Slim Jim"—a device for opening locked doors—and managed to slip it in between the window and door. The lock released easily.

Sergeant Willis scanned the map for any information that might tell them Marc's whereabouts. It was a general map of the area, but there were no marks or notes on it at all. He opened the glove compartment—just the usual—an ice scraper, some maps, car registration, and some old catsup packages from McDonald's. They checked under the seats—nothing.

"We may want to get into the trunk too, but I doubt there's anything there that will help," he said.

Sharon stood watching the snow fall and listening to the shrill cries of blue jays as Sergeant Willis called his office and put the wheels in motion for setting up a base camp here where the car had been found.

The flakes were swirling around the clearing in small gusts as two more members of the patrol arrived. It was 4 o'clock, and although it normally wasn't completely dark until about 9 o'clock this time of year, the storm clouds were already effectively blocking out the daylight.

"We'll set up the camp tonight and have searchers in the woods as soon as we can," the sergeant said. "With any luck, we'll find our man quickly."

Sharon looked at the snow now piling up in small drifts around their car and hoped that he was right.

Chapter 12

METERS

Wednesday, May 17th
1:00 AM PDST—daybreak

"Will you hand me that pair of pliers please, Marc?"
"Sure, Grandpa Bob. Think you can fix the car?"
"You bet, Marc. Just need a little piece of wire, and we'll be on our way."

"Sure is hot today, Grandpa."

"Oh not so hot, Son. Let's take a little walk down by that stream there and see what we can find."

"Okay, Grandpa."

Marc took hold of his grandfather's hand, and they left the gravel road and climbed down the embankment to the stream.

"Why there's what we need right there!" his grandfather said, pointing across the stream at a broken down fence at the edge of a pasture. "Come on, Marc, let's go wading."

The two of them slipped their shoes and socks off and waded into the icy stream. Marc held onto his grandfather's hand tightly to keep from slipping on the algae-covered rocks. On the other side, they walked gingerly into the tall meadow grass to the fence.

"Look at that, Marc. Perfect. Just what we need. If you just look around you, you can usually find what you need to fix something right in front of you. You just have to know what to look for."

His grandfather pulled out his pocketknife and wrestled the blade through a piece of the fence wire. When he had it free, he folded it in half and put it in his back pocket. Together, they waded back across the stream to their dry shoes and socks and then back to the car.

"What are you doing?" Marc asked trying to see what his grandfather was fixing with the piece of wire under the hood of the car. His grandfather had a voltage meter in one hand, and he kept looking at it and making small grunts of satisfaction.

"Well, let's just say that we're doing what your mother should have had done in a tuneup on the car. When you're a little older, I'll teach you all about that stuff. Here, take this, will you, but be careful with it. It's my most valuable tool," he said handing him the meter.

They got in the car and drove back over the country road where they had gone berry picking on the hot July afternoon.

"Mom, guess what Grandpa did to your car!" Marc yelled as he ran into the house. "Mom!" Where was Mom? This wasn't his home. He was still outside, and it wasn't where he had been with his Grandfather.

He opened his eyes and blinked.

"Oh no," he groaned. "I've been dreaming." He lay back exhausted, feeling the sweat on his face dripping down his neck into his clothes. He raised a hand to wipe his face and felt the sting from his salty sweat penetrate his cut hands.

Grandpa Bob. It had all seemed so real.

"Boy, I wish you were here," Marc whispered aloud. "Mr. Fixit"—that's what they used to call his grandfather. What was it they had written under his photo in his college album? "If necessity is the Mother of Invention, Bob is the Father."

Over the years that Marc had known him, that had truly been the case. He had taught him how to build and fix everything from transmitters to carburetors. It had been under his patient instruction that Marc had gotten his own ham license at age 14. He remembered his first QSO with his grandfather after he'd earned his license. "To KA7ITR from W6VGQ—here's to many more fine contacts."

Well, there had been a lot of "fine contacts" but not enough as far as Marc was concerned. His grandfather had died two years ago, a "Silent Key" as they listed him in *QST* and various other ham publications.

He stared up at the starlit sky. It didn't feel quite so cold tonight. He reached out and felt the grass beside him. The blades didn't feel frozen although they might be before morning. He was so hot. The coolness of the ground felt good to his hand. He grabbed a handful of snow and rubbed it against his feverish face. He lay back and closed his eyes, enjoying the feel of the night air against his wet face and soon fell back to sleep.

"Don't do that, Marc. You'll get chilled! Besides, I need you to help me to fix something."

"Okay, Grandpa Bob, I'm coming."

His parents were sitting in the living room looking at the water bill which had just arrived. $120.76!

"It says the water department thinks we have a leak," his father said. "Well so do I, but darned if I can find it."

"The water pressure has been low in the house all month," his mother added.

"Come on, Marc, let's go play detective," his grandfather said. They walked outside and looked down the long driveway to the street.

"First thing is to determine if the leak is inside the house or outside."

"How do you do that, Grandpa?"

"Easy. Just turn off the water to the house."

His grandfather walked around behind the house and shut off the master valve.

"Now, come with me."

They walked down the driveway to where the city water meter was located. His grandfather lifted the cement lid and peered down at the dial on the meter.

"Look at that, Marc. What do you see?" Marc got on his hands and knees and stuck his face close to the meter.

"It's moving, Grandpa! The dial is moving!"

"That's right, Marc, and that means that the leak is not in the house because we shut off the water to the house. Remember that Marc. Meters are tools to help you. They can help you fix all sorts of problems."

It had certainly helped them to fix this one. Marc and his father and grandfather had started walking their twenty acres of Christmas trees inch by inch. A few hours later they found the problem—a broken irrigation pipe up in the north field that had been trickling water all month.

"Grandpa, don't walk so fast. I can't keep up with you. My leg hurts."

His grandfather came back to where Marc had stopped in the field. He looked at Marc's leg but didn't seem to be too interested.

"Come on, Marc. Just forget about it. We've got work to do. Remember how we promised your parents we'd wire some new electricity outlets in the basement."

"But Grandpa! My leg hurts...!"

Marc groaned as he awakened for a minute. He had rolled onto his left side, and the twisted limb was lying underneath him.

"Ohhhhh" He struggled onto his back and lay still as the fire in his leg alternately flickered and blazed. So hot, he was so hot. And he thought his matches didn't work. He must have a fire in his sleeping bag. He flung the cover back and let the cold night air wash over him. Oh, that felt so good. Now he could sleep.

"What are you doing, Grandpa?"

"Trying to find the problem, Marc. Your mother says you keep on blowing fuses on the kitchen circuit."

"But how are you going to fix it, Grandpa?"

"Look at this little box, Marc. It's a voltage meter. See the needle there. We can measure where we have electricity and where we don't. A meter measures things, Marc. Remember that. It can be your best friend."

The two of them played sleuth all afternoon until Grandpa Bob found the problem in a faulty fluorescent light fixture.

"Want to help me fix it, Marc?"

"Oh Grandpa Bob, I'd like to, but something's wrong with my leg. I think I want to go lie down."

"No, Marc! You mustn't! You must fix the problem first!"

"Gee, Grandpa. Don't be so mad at me. You've never been mad at me before. What's wrong? Did I do something wrong?"

He started to cry, and Grandpa Bob looked troubled. He put an arm around him and gave him a quick hug, but when he spoke, he was still stern.

"Help me fix this. Don't rest until we get the job done."

"Okay, Grandpa Bob, I will. But how will we know if we fixed it right?"

"Well, if your parents don't blow anymore fuses for one thing, but even if we didn't have fuses to tell us when something was being overloaded, I could still tell with this meter. The meter, Marc. Don't forget the meter."

Marc woke up shivering. It had started to snow lightly, and the blowing flakes were drifting in and landing on him under the trees. His hot skin melted them immediately, and he was soaking.

Where's my coat? Marc wondered as he looked down at his one bare arm protruding from his torn shirt. As his eyes adjusted to the darkness, he saw his coat lying on the edge of the sleeping bag.

I wonder how I got that off, he thought, remembering how difficult it had been to put on. He grabbed it and slipped his wet arms into the sleeves. The movement woke up his leg again, and Marc felt nauseated and dizzy with the pain. Thirsty. He was so thirsty again. He reached for the canteen and shook it. Empty. He would have to wait until morning to fill it. Wincing at the pain it caused him to move his sore hands, he grabbed a handful of snow and put it in his parched mouth. He began shivering again until the warmth of sleep overtook him.

Good, Marc thought. *I'm not dreaming anymore. I'm home.* He slipped his key into the front door lock and opened it.

"Anybody home?" he yelled. There was no answer.

"I guess not," he muttered, walking into the living room. Boy, it felt good to be home. He walked over to the sliding glass windows and opened the drapes that covered them. The back yard looked so pretty with the roses and rhodies blooming.

Maybe Mom was outside somewhere. She spent a lot of time working outside.

He opened the doors and yelled, "Mom" several times but there was no answer. Oh well, she must have gone to the store, and Dad was probably at work.

He wandered around the living room, looking at things and reaching out to touch some of them. There was his Little League trophy on the bookcase with the picture of him in his uniform, freckles and all. He sure looked silly as a little kid. And there was a photo of the three of them standing in front of a cabin on the Olympic Peninsula. His dad was holding up a salmon.

Vacations. They had sure taken a lot of them together. Then he remembered. Of course. That was why they weren't here now. They were on a vacation. Well, why did he come home then if they weren't here?

Maybe his grandparents were here. They visited so often that Marc called the spare bedroom theirs. He was always happiest when they were there. But of course, they wouldn't be here if his parents weren't here. Still, it wouldn't hurt to look. He pushed open the door and stood in the doorway. The bed was made up with the blue and white quilt the way it always was. But no, they weren't there. The dresser tops were clear. When his grandparents came, his grandmother always had her overnight bag on the dresser, and his grandfather had his radio gear spread out.

He walked in the bedroom and looked at a picture of the two of them hanging on the wall. He reached out and touched it and stood looking for a long time at the two of them. He turned to leave and jumped back in surprise. His grandfather was standing in the door. He looked so sad, and he didn't say a word.

"What's happening here, Grandpa Bob?" Marc cried out. His grandfather silently reached out and took Marc's hand and led him to his own bedroom.

The door was standing open so he went in and sat down on the bed. His grandfather remained in the doorway. Everything looked so clean and neat. You could certainly tell that he was away at college. While he was here, there was a lot of stuff he'd like to get to take back, but first maybe he'd take a nap. The bed sure felt good—not like the dorm bed.

He closed his eyes and then opened them again because his grandfather was touching his shoulder. He pulled him up off the bed and led him to the closet and stood back for Marc to open it. Marc flung open the doors. It was empty! Strange. Where were all of his clothes? He tried to turn on the closet light, but it wouldn't work. Suddenly, he felt frightened of the dark closet and slammed the doors shut. He turned to question his grandfather, but he was gone.

"Grandpa Bob!" he yelled, but there was no answer.

Then he saw his desk. It was bare. All of his ham station gear was gone! What was going on here?

Frantically, he pulled open all of the drawers of his desk and his bureau. All empty. He looked out the window, craning his neck to see the antenna on the roof. Gone!

Then he saw the meter lying on his desk. Was it a voltage meter? He wasn't sure. He had never seen one like it before, but it had a scale and a hand like any meter that would read current of some kind. He grabbed it and ran into the hall calling for his grandfather. No answer.

He felt as if he couldn't breathe as the hot air in the hall swirled around him. Funny, it hadn't seemed so hot when he came in. Now, it was stifling. Well, he would get outside and cool off.

The trip down the hall was slow. Every time he tried to move his left leg, he could feel something pulling it backward. Anxiously, he turned to look, almost expecting to see some monster holding his foot, but there was nothing.

I've got to get out of here! He commanded his body to move faster as he half crawled-half walked toward the front door.

His hands hurt terribly as he reached up and tried to turn the doorknob. It wouldn't budge. Then, he looked at the meter he was still clutching in his left hand. Marc could hear the soft whoosh of the wind outside in the trees, and the meter seemed to be keeping cadence with it. Every time the wind rose, the hand on the meter swung to the right.

"Use your meter." He remembered his grandfather saying that. How was he supposed to use this meter? He was so hot—he had to escape from the house. He grabbed the door knob and tried to turn it with all of his strength. It wouldn't budge. Then he noticed that the needle on the meter was at its low ebb and was now slowly rising.

Without knowing why, Marc waited for it to peak and then twisted the knob hard to the right. The door opened easily, and he felt the cool rush of air as he tumbled outside.

Chapter 13

A NEW CODE

Wednesday, May 17th
10:00 AM PDST
1500 Zulu

G roaning and thrashing in his sleeping bag, Marc gradually fought his way to consciousness. His fever, caused by the worsening infection in his leg and cuts, had broken. As he opened his eyes to the morning fog on Wednesday, he mopped his soaking face with his sleeve. He felt faint and chilled, but for the first time in over twelve hours, his thoughts were clear.

What a dream! Or rather a series of dreams! It all seemed so real. His grandfather had been right there, and together they had been reliving all sorts of problems. Funny how all of those incidents had clumped themselves together into one long dream. And, there was a point to it too, as he remembered. Something about meters. Marc shuddered as the vision of the meter lying on his clean desk in his empty bedroom came to mind. Now, that was the one incident that had never happened—of course not—because Marc knew without any doubt that his dream trip to his home had been a vision of the future without him—what the world would be like with him dead!

The fire in his leg was starting to come alive again, and the raw oozing skin on his backside joined in the chorus of pain. As Marc surveyed his bleak situation, it seemed like death was a very real possibility. Sure, he could transmit, but he was certain it was at reduced power with his nine volt makeshift power supply. He had no idea if anyone could hear him. He felt as if he were sending messages toward faraway planets and hoping that some intelligent being would intercept them. How long could he hope to stay alive? He knew that infection was building in his body and that his rational moments were becoming farther and farther apart.

The cold wind drying the sweat on his face chilled him, and he began to shiver uncontrollably.

"If infection doesn't get me, hypothermia will," he muttered grimly between chattering teeth. "I've got to take command." The sound of his own hoarse, quivering voice scared him even more, but he shook his head and forced himself to assume a calm he didn't feel.

Slowly, he reached out to his backpack and found a package of dried soup mix. Surprisingly, he had no appetite, but he knew he must eat. He put a chunk of snow with the soup in a cup and stuck the cup in his sleeping bag to warm it up. Did he have enough strength left to throw his canteen in the stream? He tried, and the effort seemed like that for a marathon run. It hurt so much to clench his abraded hands around anything, and the throw wrenched his entire body. His first six throws fell short by at least ten feet. He alternately cursed and groaned as each throw missed its mark. Finally, he hit the stream. In the last 24 hours, the dropping temperatures had increased the ice cover near the water's edge, and, when he dragged the canteen back, it contained only a couple of inches of icy liquid.

Greedily, he drank the water and then stuffed the canteen full of snow and tucked it down into his sleeping bag. He thought the cold plastic would feel good next to his fiery leg, but the canteen "icepack" set off another chill.

He tried striking one of the matches he had stuck in his wool hat. Still damp. In fact, his sweat had permeated them. There hadn't been enough sun to dry the matchbook he had laid out on the ground and later put in his backpack.

He forced himself to eat the slushy soup mixture and then took his radio rig out of the backpack where he had put it to keep dry. He arranged the equipment on the edge of his sleeping bag and then put the backpack under his head as a pillow. He wasn't really comfortable, but the position was at least tolerable. It was impossible to lie on either side for more than a few seconds because of his leg. Lying on his sore back made him feel as if a million needles were poking into him, but at least he could operate the rig with his right hand and keep his left leg still.

Marc flipped the switch on and again breathed a sigh of relief as the S-meter jiggled in response to the power. He glanced at his watch and decided to make a schedule which he would try to keep (that is, if he could keep himself awake!). *Transmit every half hour for 3 minutes and hope that someone would eventually hear him—before it was too late,* he thought, as another chill hit him.

His hands, which had been so badly scraped in his trek down the hill, were swollen and stiff. It was an effort to bend his fingers enough to grasp the code key, but Marc gripped the key with determination.

"SOS SOS SOS SOS de KA7ITR" he sent slowly. "Injured in Jefferson Wilderness Area" He stopped trying to think of a specific location to send. Where was he? He was beginning to feel nauseated and dizzy again, and he knew from the heat in his face that his fever was coming back. He felt as if the fog blanketing the woods around him had invaded his brain. Surely, his location would occur to him if he thought hard enough about it. He squeezed his eyes shut in concentration, but the only word that came to mind was *Jefferson.*

The wind rustled the tree branches above him, making the boughs drop their snowy coverings. Impatiently, Marc brushed the snow off his face and the top of the rig, and resumed transmitting.

Something was making a vague connection in his brain. Where had he heard that soft wind sound before? Then, he remembered. In his dream. He had been standing in his hallway trying to get out the front door when he noticed that the rise and fall of the wind outside seemed to be reflected on the meter in his hand. Without thinking, he glanced at the S-meter on the front of the rig and gasped. He wasn't transmitting, but the needle was jiggling slightly! Why? Perhaps, it was registering the power surges of band noise—he doubted that it could be a specific signal.

Oh, why did he have to feel so fuzzy in his brain? He felt as if he were on the edge of understanding something very important, and yet, it was so hard to think. *Transmit*—some part of his brain told him.

Slowly, he began sending "SOS" again and then he sent "if u hr me, 5 sec resp pls" (if you hear me, 5 second response please). He held his wristwatch up before his swirling vision and then looked at the S-meter. Slowly, the needle swung to the right and stayed steady for exactly five seconds.

Frantically, Marc repeated the message. "SOS de KA7ITR if u hr me, 5 sec resp pls." Once again, the needle swung past the midpoint on the S-meter and stayed there for exactly five seconds.

Marc shouted and then wept. There was hope!

He forgot his pain as he hiked himself into a sitting position. "Pls help me," he sent slowly. "Injured and lost in Jefferson." He paused. Where was he? Somehow, the words "Vienna, Austria" popped into his mind, and without thinking, he sent them. Once again, the S-meter responded with a five second acknowledgment, but this time the needle didn't swing as far. He was losing whoever was hearing him.

Vienna! That's where my parents are on vacation! How stupid of me! I'm in the mountains, Marc thought.

"SOS de KA7ITR—lost in Jefferson Wilderness Area nr Hunt's Cove." At last, he had thought of where he was, but it was too late. There was no answering response on the S-meter. He transmitted again, making sure that the meter reflected that he was still getting out. No telling how long his battery arrangement would work. Yes, he had power, but again no response. Should he waste battery time by continuing to transmit or should he wait? Reluctantly, he shut off the rig and looked at his watch. He would wait another half hour. Maybe band conditions would be better by then.

The moment his attention was off his QSO, the pain took over. Groaning, he edged himself down farther in the sleeping bag so that his weight was distributed along his entire body instead of concentrated on his seat. It was almost noon, and he could feel the temperature in his body starting to rise. He remembered now that was how it had happened yesterday. He had been fine in the morning, but the afternoon and night were lost to fever and wild dreams.

Anxiously, he looked down into the sleeping bag at his leg. The grotesque swelling was still there, and angry looking red streaks were starting to extend from his knee up his thigh. He tried to wiggle his toes, but found that it was impossible. What scared him even more was that it was becoming hard to move the toes on his right foot as well. Frantically, he moved his right leg around in the sleeping bag as much as he could. He wished he could reach down and rub his right foot which felt cold, but the effort of bending down that far sent excruciating arrows through his left leg. So he was getting frostbite, he thought—on top of everything else. And what about the red streaks? He vaguely remembered an uncle who'd had blood poisoning—remembered the family talking about "red streaks," and how massive antibiotics had saved him "just in time."

"Hurry, please someone hurry," he whispered. Forget the schedule. He was going to transmit now! He pulled the code key closer to him and began frantically sending. While the signal strength meter dutifully showed his own transmission, it lay still the rest of the time. No one was hearing him. Maybe his batteries were going dead. Maybe he had just thought he had seen a response earlier. Maybe...he could feel hysteria rising in him and then lulling as a tremendous tiredness washed over him. He closed his eyes and dozed.

Marc had no real recollection of the afternoon. He was awake part of the time, but everything seemed to be happening in slow motion—like a dream in which he was trying to run and could only move at a snail's pace. He was alternately hot and cold. Sometimes when he woke up, he had flung the sleeping bag off. Other times he woke to intense shivering and pulled his cover up as tightly as he could. Each time he awoke, he thought vaguely of transmitting, but his hand never seemed to be able to reach out and grab the key. He slept—no dreams—just unpleasant hot and cold sensations that carried him through a dark void.

At about 5 o'clock in the afternoon, he woke up to a swirling snowstorm. Everything was rapidly becoming blanketed under the white fury. Then for just an instant, the clouds parted, and he noticed the sun low in the sky.

"It's morning," he whispered. "I've slept through another night." He lay still, paralyzed by the idea that time and life were slipping away from him. As he watched the blowing snow, he felt peaceful. Perhaps, he would just go to sleep and not wake up. Then the fire would go out in his leg. And in his life, too.

Suddenly, the trees shook as a strong wind blew gusts of snow through them. The sun peeked around the edge of a dark cloud, and with surprise, Marc noted that it was lower in the sky.

"It's not morning—it's afternoon!" he cried almost joyfully. "It must still be Wednesday!"

It was as if someone had given him an extra day of life. He was relieved to know that he could still keep track of the time.

He brushed the snow off his equipment and began to transmit. Over and over, he sent the same message.

"SOS SOS SOS SOS de KA7ITR injured in Jefferson Wilderness."

There was no response. Could he have been imagining the earlier one? He stared intently at the S-meter and thought he saw it jiggle a little. Once again, he sent his message, and once again the meter responded faintly. If another station had asked him how he was copying them, Marc would have said "very weakly—about an S2" according to the indication on the S-meter. He knew that the other party was probably copying him about the same way—if at all. With an S2 rating (on a scale of 9) it was doubtful if they could make out what Marc was sending. Still, he persisted, and each time he saw a faint response on the meter.

Finally, he was afraid he was wasting battery power without a solid contact, so reluctantly, he turned off the rig. The water in his canteen was partially melted, and he took a drink, letting the lumps of snow slide down his parched throat.

It was dusk and Marc knew he should eat. Food was in his backpack right behind him, but he couldn't seem to summon the strength to turn around and open it. He was so tired. Just a little nap, he promised himself, laying his head back on the pack. Sleep instantly took him, and for the next hour he journeyed painfully through heat, cold, pain, and disconnected nightmares. He was aware that he was dreaming, and kept reminding himself that he needed to wake up—that he had a schedule to keep. He felt himself take command of his sleeping body and order it to become conscious. Slowly, he opened his eyes and stared out at the white valley.

Why couldn't he see Mt. Jefferson? Suddenly, he knew. *I haven't crossed the Crest Trail yet—that's why,* he realized with amazing clarity. *I know where I am!* Marc thought. *I'm a little past Hunt's on the way to Coyote Lake.* He grabbed the key to send this new information and then froze. A strange rich musky odor assailed his nose, and he felt the hair on his neck involuntarily bristle. He knew even before he looked up that something was watching him. Slowly, he raised his eyes and found his stare deadlocked with that of a large black bear ten yards away. Her frolicking cub was ambling through the drifting snow toward him, and the mother raised herself to a standing position and growled a warning.

Chapter 14

BASE CAMP

Wednesday, May 17th
3:00 PM PDST
1500 Hours Military Time

K im's mother barely had time for a "hi" to escape her lips as Kim came whirling through the front door.
"Have you heard anything?" Kim asked breathlessly.

"Not a word about Marc, Honey. But, Mike called from Oregon State, and said he was going to drive up here after his last class. He'll probably be here any minute. Why don't you call the Sheriff's Office now? We told them we would check back this afternoon."

Kim dropped her books in a heap on the dining room table and rushed to the phone in the kitchen. It had been such a long day at school. Normally, an excellent student, she had kept her eyes riveted on the clock, willing it to advance. More than one teacher had asked her a question and caught her unprepared. Now that she was home, the school day seemed like a blurry dream to her. She quickly dialed the Sheriff's Office.

"Marion County Sheriff's Department," the dispatcher answered.

"Hello, this is Kim Stafford. I'm the Amateur Radio operator who reported Marc Lawrence as missing. Have there been any developments in that case?"

"Just a minute, Miss Stafford. I believe there have been. Lieutenant Baxter, who's in charge of Search and Rescue, said he wanted to talk to you."

There was a pause and then a deep male voice came on the line.

"Lieutenant Baxter here."

Kim identified herself, trying to keep the anxiety she was feeling out of her voice.

"Miss Stafford, we do have some news. We found Marc Lawrence's car this afternoon, and we're in the process now of setting up a Base Camp near that area. We're gathering together all the search and rescue units, and they should be deployed soon."

Kim paused, catching her breath.

"Could I go up there? To the Base Camp, I mean?"

"Yes, you can, but I want to caution you that it has begun to snow. Do you have a vehicle with traction devices?"

Kim heard her mother letting Mike in the front door, and she waved to him as he came down the hall.

"Just a minute, Lieutenant."

"Mike, what kind of car do you have?"

"It's a Toyota four-wheel drive pickup—why?"

"Want to go to the Base Camp? They've found Marc's car."

"You bet—let's go."

"Yes, Lieutenant. Marc's roommate from college is here, and he has a four-wheel drive. He'll come with me."

She listened while Lieutenant Baxter gave her careful instructions to the Base Camp. When she hung up, she saw that her mother had fixed a thermos of hot coffee and some sandwiches.

"No telling how long you kids will be up there. Do me a favor though, and call me when you get there so I know you made it okay."

"Sure, Mom," Kim said, giving her a hug.

Kim grabbed her two-meter rig and an extra battery pack and her winter coat. Within minutes, she and Mike were on their way up Highway 22 to Detroit Lake and the Base Camp on Road 2246.

The rain was coming down heavily, and the two of them rode silently with only the swishing cadence of the windshield wipers as conversation. Finally, Kim broke away from her own thoughts and turned to Mike.

"So, how was your day?" she asked.

He laughed at her bland question, then answered it truthfully.

"Awful. I couldn't concentrate at all. I just kept thinking about what it would be like to be freezing somewhere in the woods."

"Me too," Kim said, shuddering. "Mike, tell me more about Marc. What's he like? Do you think he's the type who could survive in the wilderness?"

"I don't know, Kim. How can I answer a question like that? He's a very resourceful guy, and I think he's pretty tough, but I suppose it would depend on how bad a situation he's in."

"Yeah," said Kim softly. "Of course. That was a stupid question to ask."

Mike reached out and squeezed her shoulder lightly.

As soon as they passed Mill City, the snowflakes began mingling with the raindrops on their windshield. The road was clear though, and it wasn't until they reached Detroit that Mike had to switch into four-wheel drive.

Kim asked Mike to stop a minute while she ran into a market to call her mother.

"They were talking about the search in the store," she said as she got back into Mike's truck. "The clerks have seen all the sheriff's vehicles going by, and they're wondering what's going on. One of them said something about a murder!"

"Did you tell them what's really happening?"

"Nope, didn't want to take the time. Let's go."

Road 2246 was easy to spot. A member of the Jeep Patrol was parked on the shoulder and standing by his truck. Mike and Kim waited while he gave directions to the driver of a truck pulling a horse trailer with a small emblem on the side saying "Oregon Mounted Posse."

Mike rolled down his window and identified himself to the Jeep Patrol man.

"The camp's about three miles down this road. Just go slow—it's pretty slippery."

They followed the horse trailer down the road. Kim watched the horses struggle for balance, their tails swishing impatiently over the tailgates, every time the trailer hit a bump in the road. *They're probably used to this,* she thought. *Certainly a lot more so than I am,* Kim said to herself as she dug her fingernails into her palms.

They came into the clearing at the end of the road and parked under some trees out of the way of incoming traffic. Kim was amazed at all the activity. Several sheriff's vehicles were there, and a sheriff's officer—who she assumed might be Sergeant Willis— was directing everyone.

As a man unloaded two chestnut-brown quarter horses from the van, Kim overheard Sergeant Willis tell him that the rest of the group wouldn't be there until morning.

An eighteen-foot trailer owned by the local React group was in place, and someone inside was busy cooking. Kim could smell the faint aroma of stew as she and Mike got out of the truck.

Just then a large communications van came lumbering down the road, and Sergeant Willis waved it over to a clear spot near the trees. A man and a woman got out and immediately began setting up a large light pole, powered by an emergency generator.

The snow was coming down in blowing sheets, and through the whiteness, Kim could barely make out the outline of Marc's car at the far end of the parking area. Someone had cordoned off the area with ropes, and the car sat isolated beyond the bustling activity of the camp. Snow was rapidly covering the car, but the ham antenna on top of the roof was clearly visible.

Mike caught the direction of her gaze and said, "That's his all right."

The two of them turned as Sergeant Willis came up to greet them. He was a pleasant-looking man, probably in his early forties, Kim thought, with the look of both authority and kindness in his face.

"I bet you're Kim," he said, extending his hand. "Lieutenant Baxter radioed me that you were coming. And, you must be Mike," he said, shaking his hand too.

Kim felt somehow comforted by being called by her first name—everyone had seemed so formal on the phone.

"Boy, you guys really mean business," Mike said, looking at all the activity.

"We sure do," Sergeant Willis said. "This is standard procedure for any search, and this isn't everyone of our group. We have Horse Posse members who will start to search from Breitenbush, working south to here, and from Linn County, moving north, at daybreak tomorrow. They'll rendezvous with the Posse members who will leave from here.

"The Explorer Post members will be here early in the morning to search the areas off the trails. Jeep Patrol is already driving all the nearby roads to see if Marc possibly hiked out in another direction."

"Can we help?" Kim asked.

"Well, not yet, but you might be able to, later. I see you brought your two-meter radio. Do you think Marc might be transmitting on two meters?"

"I talked to him on 80 meters, but the description Mike gave of his gear makes me believe that he might have also taken a two-meter rig with him," Kim said. "I tried to break the local repeaters on the way up here, but I couldn't."

"That's what I've heard," Sergeant Willis said.

"Before I left Salem, I called a ham friend and asked her to put out a bulletin on the ARES (Amateur Radio Emergency Service) net tonight for people to listen for him on both two meters and 80. If he can be heard locally, I'm sure someone will catch him. And while I'm here, I can listen on simplex, just in case he tries that," Kim suggested.

"Good idea. You two feel free to look around. You'd probably be especially interested in the communications van when they get set up, and that's where you'd hear reports of anything happening. Besides, it's warm in there," the sergeant said pulling his coat collar up around his neck. "We'll have some stew ready before long. Let me know if you have any questions."

They thanked him and walked around the camp for the next half hour, impressed by the activity going on. When it looked like the communications van was in operation, they went over and knocked softly on the metal door.

"Come on in," a friendly female voice said.

The two of them opened the door and stepped inside the compact trailer. A young woman, in her twenties, sat at a long wooden table filled with radio gear which was all securely bolted down.

"Hi. We just now got set up. Unfortunately, we've had to do this so many times, that we're getting pretty fast at it. I bet you're the Amateur Radio operator I've heard about," she said, turning to Kim. "By the way, my name's Mary Hammond."

Kim and Mike introduced themselves and asked a few questions about the radio gear in the trailer.

"Well, we're set up to monitor and communicate on all the search and rescue frequencies as well as the sheriff's bands. Nothing with ham radio, though, because I don't have a license. A couple of our Jeep Patrol members are hams, but they transmit back to us on the sheriff's frequencies. I'd like to get a ham license myself, one of these days. Maybe you can tell me a little more about it while you're here," said Mary.

"You bet," promised Kim, glad for any opportunity to talk about her favorite hobby, even though now it was difficult to think about anything other than Marc's plight.

Just then, a call came in from a Jeep Patrol on a perimeter road of the wilderness area. After she was done acknowledging their transmission, Mary talked to Kim and Mike.

"Things will really be busy tomorrow. The Posse members and Explorer teams all carry radios too. They'll be checking in every half hour tomorrow. Right now, I would guess they are our best bet for finding Marc quickly, since the weather's too thick to spot him from the air."

There was a knock, and a stocky gray-haired man opened the door.

"Soup's on, or rather I should say stew's on!" he said, bearing three steaming bowls of thick, meaty stew.

Despite her worry, Kim found that she was suddenly ravenous.

"I just remembered, I forgot to eat lunch," she said smiling.

"Well, this ought to take care of that," the cook said, handing her the first bowl.

Mike, Kim, and Mary ate while they listened to various conversations on the air. Nothing much was happening—just groups reporting in as required.

When they were done eating, they carried their bowls back to the kitchen tent. Sergeant Willis was helping a couple of volunteers start a small bonfire, but when he spotted Kim and Mike, he came over to join them.

"We finally reached the last of our Horse Posse. One of the guys is flying in from Arizona tonight, but they'll all be on the trails early in the morning. The Jeep Patrols are driving all the roads. So now comes the hardest part—we just wait and hope and pray," he said glancing at his watch. "Do you kids plan to stay up here or what? If you want to go home, I promise you we'll let you know the minute anything happens."

Kim turned to look at Mike, and he shrugged his shoulders.

"It's up to you, Kim. I'm game for anything."

Suddenly, the door of the communications van flew open, and Mary yelled excitedly, "I've got some news!"

The three of them ran to the van.

"Lieutenant Baxter just called me, and he said that a ham on the coast—let's see, " Mary said looking at her notes, "a KD7YB in Tillamook—anyway, she heard a distress call from KA7ITR. He was asking anyone who heard him to send a five second response," Mary said looking puzzled. "And the weirdest thing is that first he sent his location as Jefferson Wilderness—that's how she knew to call Marion County—but then in the next transmission he said he was in Vienna, Austria!"

"That's our boy!" Sergeant Willis exclaimed. "His parents are vacationing in Vienna—I bet he's hallucinating."

"Does that mean he's got hypothermia?" Kim asked, feeling her heart pound.

"I'm not sure, Kim. But I do think it means we have to hurry," he said softly, looking at the snow falling steadily from the darkening sky.

They sat anxiously for an hour or so, listening to radio reports, but nothing more came in. The Jeep Patrols called in that they were about ready to quit for the night. Nothing had been found on any more of the roads.

Kim sat, clenching and unclenching her hands. Finally, she stood up.

"Mike, I want to go back home—please! I want to listen for Marc on the air. I heard him once—maybe I can hear him again."

There was no denying the urgency in her voice, and Mike quickly agreed. They walked over and told Sergeant Willis that they were leaving. He cautioned them about driving carefully and promised to call them if anything developed.

"I plan to come back in the morning," Kim said, "but right now, I think the best help would be if I could hear him."

"You might consider getting a little sleep, too, young lady," Sergeant Willis said in a fatherly tone.

Kim smiled but didn't say anything.

Mike and Kim crunched through the fresh snow to Mike's truck and carefully turned the vehicle around to make their way down the access road.

As they reached the main highway, the snow stopped falling, and the clearing clouds gave them a glimpse of the starlit sky above.

"Any other time, I would say 'what a beautiful night' but all I can think of now is how cold it must be for Marc," Mike said.

"I know," Kim whispered as flakes began falling again.

The snow turned into gentle rain long before they reached the foothills surrounding the valley. The porch light was still on at Kim's home.

"What do you want to do, Mike? You're welcome to come in and spend the night—we've got a guest bedroom."

"Actually, I think I ought to go back to school. I have a big exam at eight tomorrow morning—not that I feel much like taking it, but I guess I ought to try. Will you promise to call me any hour of the night if you contact him or if you hear anything more from the Sheriff?"

"Sure, Mike. I will."

"I'll be free after that test so I could be back here by 10 o'clock. Want me to take you back up the hill?"

"Well that would be great if you could. Otherwise, I can put chains on the VW."

"I'll be here," Mike said emphatically. "Remember, now, call me if you hear *anything*."

Kim gave him a quick hug and ran into the house. Her parents were watching TV in the living room. She told them what had happened and explained that she was going to listen for Marc.

"And, please don't bug me if my light's on all night. I can sleep some other time in my life," she begged.

"Don't worry about that, honey," her father promised. "You just let us know if we can help."

Kim ran up the stairs and sat down at her rig. Carefully, she tuned up her transmitter on 80 meters.

"Be there Marc—oh, please be there, Marc!" she whispered.

Chapter 15

— • — •• — — K I M

Wednesday, May 17th
7:00 PM PDST
0200 Zulu

Sweat slowly dripped down Marc's forehead as his gaze remained locked with that of the mother bear. It had probably just been seconds, but he felt as if he had been transfixed like this for hours. Slowly, he let his breath out and tried to take another without any noise or movement. The mother bear remained standing, her head held high, sniffing for his scent. From the corner of his eye, he could see the roly-poly baby cub frolicking toward him, ignoring his mother's growled warnings.

"Go away," Marc prayed silently. The cold sweat trickling down the back of his neck made him itch. The mother bear opened her mouth wide, and Marc saw the formidable white teeth. He tried not to shudder.

The cub stopped and raised his head in the air, closely imitating his mother. His shiny, black fur rippled over his muscular, fat, young body as he came closer and closer to Marc.

It was almost dark, and the wind shook clumps of snow from the trees. A blue jay made a noisy protest as falling snow from a branch above knocked him off his perch and sent him searching for a new place farther up in the tree. The bear cub turned his head at the commotion the bird made and caught a glimpse of the bright blue tail flashing through the branches. Intrigued by this new diversion, he turned and scrambled as fast as his baby legs would let him toward the tree which was close to his mother.

Still sniffing the air, she came down on all fours and growled what sounded like a lecture to her baby as he stood

at the base of the tree looking up. She nuzzled him and, satisfied that no harm had befallen him, she growled again, and the two of them disappeared into the woods.

Marc sat motionless, watching the spot where they had made their exit. He could feel his whole body beginning to shiver, and he didn't know if it was from the fever or fear. The shivering intensified until even his legs and arms were jumping and his teeth were chattering. Every shaking movement sent a jolt of agony through his leg, and he grabbed it with both hands to try and keep it still.

And then just as quickly as the shaking had gripped him, warmth was spreading through his body even before the shivering stopped. The chills and fever were like two trains speeding in opposite directions, and for a few brief seconds when they passed, Marc felt neither too hot nor too cold. Then the fever train took over, and he felt his temperature soaring.

He was so thirsty! He reached out to his canteen and groaned when he found it empty. The line was still attached to it, and he tried tossing it into the stream, but even his best effort fell way short. He lay back, panting, on his backpack, and then stuffed a handful of snow into his mouth. It felt so good sliding down his hot throat. He grabbed some more and savored the feel of the icy crystals on his tongue.

He knew that eating snow in his weakened condition would risk lowering his body temperature even more, but his thirst was too powerful to ignore. He kept putting more clumps of snow into his mouth until his throat no longer felt parched, and then he stuffed his canteen full and put it under his sleeping bag to warm.

Now the fever train had reached the end of its track, and he was sliding backward toward cold and shivering again. Desperately, he reached around to the pocket on his backpack and pulled out its contents. His German dictionary and a book of matches were the first to hit the ground. He held the matches up to his cheek and thought they seemed pretty dry.

A fire! If only he could build a fire! It would signal help, keep away bears, and best of all, keep him warm. All of the branches on the ground next to him were covered with snow.

He dug frantically under them down to the ground and succeeded in coming up with a few fairly dry twigs.

"You should see me now, Dr. Keller," he said as he ripped pages from his textbook and crumpled them. Somehow, he thought his German teacher might approve—he was certainly putting the book to good use.

Carefully, he placed the twigs on top of the paper, and then, shielding the match from the wind with his hands, he tentatively struck it. The first one was limp and refused even to spark. He felt all of the matches and came up with two that seemed dry.

The first one lit with a tentative flicker and Marc touched it to the dry paper, moving his body to shield the flame from the wind. The tiny fire licked along the edges of the pages as Marc blew gently on it trying to encourage the wispy flame.

A small crackling noise cheerfully broke the silence as the first twig ignited, and Marc felt a ray of hope as the fire began to grow. He pulled his backpack over to act as a windbreak for one side and placed his left hand on the other side. Gradually, the flames began to take hold, and soon Marc had enough of a fire to warm his hands.

The heat made the infected cuts burn and itch. He wasn't sure the warmth was an improvement over the numbing cold, but at least it was a little easier to bend his fingers.

Fuel! He must have fuel to keep the fire going. He reached under his sleeping bag as far as he could in both directions and pulled out the few twigs that were there. The branches elevating his leg were fairly dry too, and he broke off the smaller side branches to use. He thought of pulling all the wood out from under his leg to burn, but he decided that brief warmth was not an adequate trade-off for the torture involved in having his leg lie flat.

As the night deepened, Marc's obsession with the fire grew. He pulled out his pocketknife and dug some of the bark off the tree behind him. Because his leg was so tender and swollen, he could no longer roll from side to side. His reach was restricted to what he could grasp from where he lay on his back.

He reached out and made sweeping gestures with his arms to clear the snow in search of dry wood. The movement of his arms reminded him of making snow angels as a child. The minute the thought popped into his mind, he knew it was a mistake to let his brain wander from the task at hand for even a second. His fever was soaring, and as he lay there, his eyes closed. He felt himself drifting helplessly through delirium. He felt as though he had become divided. There was still a small part of him that was conscious; it was urging the floating part to come back. Like two warriors, the halves fought, and finally with a start, he opened his eyes. He was whole again, and his hand was clasped around some small twigs under the snow.

Carefully, he fed them into the fire and added a few more German pages to ensure that they burned. With fascination, he watched the flickering flame. It was a mixture of red and white and yellow, and when printed pages were added, a brilliant blue joined in. He held his snow-wet hands out to the heat and tried to rub them together. The friction hurt too much, so he contented himself by just slowly clenching and unclenching his stiff fingers.

The flame was hypnotizing. Marc put his backpack behind him again so that he could lie back and watch the flame. *Amazing!* he thought. *I can see the fire there and yet I can feel it in my leg. It's become part of my body.* Soon the fire had become his body, and he flung back the sleeping bag as heavy sweat drenched him. The part of him that had been rational was lost to the hallucinations that followed, and Marc alternately slept and talked nonsense aloud.

Around 4:00 AM, he awoke with a start to find that his fire had smoldered away and was now covered with snow.

"No!" he shouted when he realized that he had been asleep for several hours. Not only had he let the fire go out, he had forgotten to keep his transmitting schedule.

He lay as still as he could with the shivering that had begun again and watched the sky again for the first sign of light. Without a flashlight, he needed at least some daylight to be able to see the S-meter on the transmitter.

Soon, the whiteness of the snow reflected the first morning light through the clouds, and Marc painfully raised himself into a sitting position to transmit.

He pulled the transmitter, tuner, battery pack, and code key all into his lap so that he wouldn't have to turn. He was thankful to see that his antenna wire still seemed to be in place. He was afraid that snow might have knocked it down.

Marc patted the transmitter, like a genie rubbing a lamp, before he turned it on. He switched it on and watched the S-meter respond to the power surge.

His right hand was so stiff he couldn't grasp the key. Finally, with his index and middle finger held awkwardly together, he managed to slowly punch out his code message.

"SOS SOS SOS SOS SOS de KA7ITR."

He sent his message several times and then stared intently at the S-meter. It was jiggling rhythmically.

Marc rubbed his eyes, which felt blurry. What was it he had sent before? Thinking was like running through deep water. Every movement was such an effort. Slowly, it came to him.

"If u hr me 5 sec resp pls." He hammered the message out, wincing each time his fingers pushed down on the key.

The S-meter responded with a steady five second swing, and then it began jiggling again. Fascinated, Marc watched it. There seemed to be some sort of pattern to it, but he wasn't sure. The person was sending very slowly.

Marc bent over and pulled the transmitter as close to his eyes as he could. The jiggling was continuing. Without thinking, Marc began to repeat aloud what he was seeing on the meter.

"Dah dit dah, dit dit, dah dah." Over and over, the pattern repeated itself on the meter. The pain and fire in Marc's body was so intense that it took all of his concentration to watch the meter. He wanted to scream, but instead, he clenched his teeth and whispered the code through his lips. "Dah dit dah, dit dit, dah dah." What on earth was that?

He shook his head. "Dah dit dah — • — that's K!" he almost shouted. "Dit dit! •• I Dah dah — — M! Kim!" Saying

113

the name aloud sent a feeling of joy through him, but in his foggy state, he wasn't quite sure why. "Who is Kim? Come on Marc," he asked himself fiercely. "Who is Kim?"

With a flash, he knew, and the fear he felt when he realized that his mind was not working well, made him tremble with an intensity that matched his returning chills.

The meter stopped, and Marc gripped the key with determination.

"• — • R • — • R" he sent over and over meaning "Roger—he had received the message." His vision was clouding over, and he felt nauseated. He knew he was about to pass out.

"— • — •• — — Kim" he sent several times. "Marc hurt. Pls hurry." Then his hand fell off the key, and he slipped back into the dark forest which welcomed him. The S-meter continued to jiggle a message. "QTH?" Marc lay unconscious, his batteries and life slowly ebbing away.

Chapter 16

CONTACT!

Thursday, May 18th
5:30 AM PDST
1230 Zulu

I n the early dawn hours on Thursday, May 18th, the Stafford house lay in quiet slumber as Mr. and Mrs. Stafford and Brandon enjoyed their last moments of sleep before starting the day. However, inside Kim's room, it was anything but peaceful.

She sat at her ham rig, frantically sending Morse code and then tuning the dial to listen. Nothing. She had lost him. She pushed her chair back and rubbed her eyes wearily.

5:30 AM, or 1230 Zulu she noted on her clock. It felt like ten nights instead of just one that she had sat glued to her rig. After Mike had dropped her off the night before, she had immediately rushed to her room to see if she could hear Marc on the air. The information that he had been heard by another ham gave her new hope. She wasn't sure why she thought she could talk to him more successfully than the other ham had, but she felt determined to try.

Her mother looked in on her about midnight and then went to bed. Kim sat transfixed with her headphones on, tuning the dial slowly. Any bit of static or faint signal got her attention immediately. She tried transmitting to Marc— "KA7ITR de KA7SJP" but only silence answered her calls.

In the long night hours, Kim mulled over what the operator at the base camp reported Marc had transmitted to the Amateur Radio operator in Tillamook. "If you hear me, 5 second response please." Why would he be sending that? she wondered.

She scanned through her notes on her desk from the long QSO she'd had with Marc on Saturday night. That seemed ages ago, now. When she was copying code, she scribbled rapidly and messily on any paper handy. Now, she gathered together all the sheets still strewn over her desk and began to read them.

In the second transmission, he had sent information about the type of rig he was using—a Heathkit HW-9. Kim pivoted her chair and grabbed a stack of magazines and catalogues from the bottom shelf of her bookcase. Sure enough, there was a Heathkit catalog among them, and she pulled it out and opened it on her desk. It didn't take her long to locate the HW-9 transceiver on page 37.

Kim studied it carefully. It was a small dark grey box with a tuning dial and a signal strength meter showing on the front. There didn't appear to be an external speaker.

"Probably using headphones just like I am now," Kim whispered aloud.

What if something had happened to those headphones? Suddenly, Kim understood. She began tuning the dial with renewed energy. She must find Marc! She called him repeatedly, wishing her own energy could become part of her radio signal. Nothing.

The night drifted on. Daisy, Kim's cat, startled her as she came into Kim's room and jumped into her lap. Grateful for the company, Kim petted her until she settled down cozily in the chair with her. Despite her worry over Marc, she was beginning to feel tired, and she could feel her eyelids drooping as she concentrated on listening to the 80-meter band. Occasionally, a strong signal would startle Kim, and she would tune slightly away from it, hoping it wasn't covering up an attempt by Marc.

At 6:00 AM, Kim was considering turning the rig off for a few minutes to go make a quick cup of hot chocolate. She felt stiff and cold from sitting in her chair all night. But every time she reached out to turn off the switch, she pulled her hand back and promised herself "just five more minutes."

She tuned the band slowly, willing her ears to be keener than they had ever been. There was a faint signal near the frequency where she had first contacted Marc. Frantically, she tuned it in, raising the volume as high as it would go. Then she gasped. "SOS SOS SOS de KA7ITR" It was Marc!

The signal had a strange chirping quality to it that she hadn't remembered from her original QSO with Marc. Had something happened to his equipment? Kim pondered this and other questions as she copied down his message.

"If u hr me, 5 sec resp pls." Just as the other ham had reported. Over and over he sent his call letters and then the strange message. Except, it didn't seem strange to Kim now.

Quickly, she responded with a five-second carrier. Then she leaned forward and fixed her eyes on her own signal strength meter as she began sending "K — • — I •• M — —" over and over. She watched as her own meter bounced in rhythm to her sending. A little too fast. She had to make sure that anyone watching a signal strength meter would interpret the bouncing as code.

"Please see it, Marc," she said.

His message began once again. "KA7ITR if u hr me 5 sec resp pls."

Again, she sent him a five-second carrier response, and again she sent KIM over and over slowly.

There was a pause and then Kim heard the joyful noise of "r" several times in her earphones. Roger! He heard her! There was more—his code sounded sloppy now. He was sending a letter and then a series of dots meaning he had made an error. Kim wrote frantically as he tried to complete his message. Finally, it came through. "hurt pls hurry."

Kim sent a long series of "r's" to indicate she had understood and then paused. What else could she send clearly enough to register on the meter. She thought of " Where r u?" but decided it was too complicated. Finally, she decided on "QTH?"—the Amateur Radio operator's Q signal for location.

"— — • — Q — T •••• H •• — — ••?" She repeated it several times, watching her own meter bounce in rhythmic response. Then she switched back to receive. Silence.

"Marc!" she cried as she began transmitting again. She tried another series of "r's " and then her name again. "KIM — • — •• — —" Silence.

Vaguely, she could hear her family beginning to stir as she alternately sent and listened. She glanced at her clock—6:30 AM. Almost a half hour since she had first heard him.

Kim turned as her mother appeared in her bedroom door. Without taking off her headphones, she quickly explained what had happened.

"I heard him, Mom, and he heard me! He's hurt and he said to hurry. Then he disappeared. Oh, I've just got to find out where he is. His code is all garbled. I think he must be in terrible trouble. He could be dying!"

Kim didn't even realize she was crying until her mother put her arm around her and gently wiped away her tears.

"Why don't you go take a hot shower and have some breakfast—you've been up all night."

As Kim violently shook her head "no," her mother said, "Listen, call someone else in your ham club and have that person listen for a while. You've just got to have some rest."

Kim still refused until her mother made the comment "You know, you may be needed a lot more later on—don't use all of your energy now."

Reluctantly, Kim called Pete, a ham friend who was retired. She knew he was an early riser. Quickly, she explained the situation. Pete had already heard about it on the ARES net the night before, and he readily agreed to cover 80 meters for her for a couple of hours.

"Promise me, you'll call if you hear him," Kim begged.

"Don't you worry about a thing, Kim. You'll be the first to know."

Kim called the Sheriff's Office. Lieutenant Baxter was in a meeting, but she left a message for him that she had heard Marc. Then, she took a long hot shower and went downstairs to breakfast.

"I'm not going to school, and that's final," she said to her parents. Neither one of them argued. Kim's mother filled up her cup with hot chocolate again and patted her on the back.

Kim was back at her rig long before Mike came at 10 o'clock. Her mother had left for work, and Mike had to ring the doorbell several times before Kim heard him with her earphones on. She ran to let him in.

"I heard him!" she told him excitedly as he followed her back to the rig. With a rush of words, she told him about the QSO.

"Guess I don't need these anymore," she said, laying aside the headphones and turning on the external speaker. "I feel like I can listen more carefully with headphones on—blocks out all the house noise, so I usually use them."

"Go ahead," said Mike. "Don't stop for me."

"No, this is fine. Besides, I was just about to call Pete, another ham who's been helping me, and ask him to take over. Do you want to go back to Base Camp?"

"Well, sure. I'd like to be there if they find him."

"Let's go then. I'd like to monitor 2-meter simplex at the Base Camp," said Kim. "I think there's a chance Marc might try that. If nothing has happened by tonight, I'll probably come back home and listen for him. Every contact I've had with him has been at night."

The two of them left and drove once again up the scenic highway toward the Jefferson Wilderness Area. It had snowed more in the night, and plows had pushed the accumulation to the sides of the highway, forming small drifts. A few snowflakes were falling from the low grey sky as they neared the Detroit Lake area.

There was no trouble spotting the road head. Two local television station trucks were stopped on the shoulder while their drivers talked to the Jeep Patrol person parked there.

"Wow," whistled Mike. "Looks like this is getting big. Wonder how they found out?"

"Oh, I imagine, the Sheriff's Office sends the media a news release every day. Sometimes, radio, TV, and newspaper people monitor the law enforcement frequencies so they know when a story is breaking," Kim said.

The Jeep Patrol member recognized Kim and Mike and waved them on. Apparently, the TV trucks got permission, too, because soon they were lumbering down the road after them. Sergeant Willis came out of the communications van to greet the approaching caravan. He waved to Mike and Kim and pointed to a place for them to park. Then, he went over to talk to the news people.

Kim and Mike walked over to the communications van and knocked. Mary's friendly voice immediately invited them in. A stack of used coffee cups attested to what kind of night she'd had.

"Any news?" Kim asked anxiously.

"Yes, there has been. Three other Amateur Radio operators have called in to Marion County Sheriff's Office." Mary glanced at her notes. "One from Vancouver, Washington, one from Olympia, Washington, and one from Bend, Oregon. All three reported hearing KA7ITR this morning at approximately 6:00 AM, and they heard your response to him."

"Yes," said Kim, nodding. "That's when I talked to him, but then I lost him. Has anyone heard him any other time?"

"No, I'm afraid not, Kim. But, I guess there must be plenty of people listening now. Someone has done a good job of spreading the word."

Kim thought gratefully of the ARES net—probably, the message had been communicated to other nets on the West Coast.

"Have your searchers found anything?" she asked hopefully.

"Nothing so far," Mary said, pointing to a detailed map of the wilderness area pinned on the wall. "The Explorers left at daybreak, and they're searching the areas off the trails in this area for any sign. So far, there's been nothing. I guess the trails were dry when Marc walked in and now they're covered with snow—makes it difficult to find any tracks."

She turned her attention back to the radio as all three of the Posses checked in. The one that had left Base Camp just finished circling Pamelia Lake and now was starting to head up the steep switchback trail toward the Crest Trail. The other two were approaching from the north and the south and would eventually rendezvous with the third Posse. No one had seen anything that looked as if a hiker had passed by.

120

The door opened, and Sergeant Willis poked his head in just as the Posse that had started from the Big Meadows area south of them checked in.

"That puts them about here," he said, pointing to a spot on the map. "After they all meet, they'll take different trails back out to make sure everything has been covered."

He turned and looked at Kim as if trying to decide how well she was holding up before making his next statement.

"Kim, as you may have guessed, the search has been picked up by the local TV stations. They want to interview you, if you don't mind. You're the local heroine," he said, smiling.

"No one's a hero until Marc is found," Kim said emphatically, "but I'll talk to them if you want. Maybe some news coverage will alert even more ham operators to listen for him."

Sergeant Willis opened the door for her and escorted her over to the TV trucks where crews were busy setting up cameras. An attractive young woman reporter, bundled up in a heavy coat and muffler, came over to interview her.

Kim had a brief thought about how bedraggled and tired she must look after her sleepless night, but the reporter was very congenial and soon put her at ease.

"I'm just going to ask you a few questions on camera, and I want you to tell how you got involved in this—okay?"

Kim nodded and then squinted as someone shined a light in her face.

Then the videotape was rolling and she found herself talking quite naturally to the reporter. The story of how she had heard Marc on Saturday night came tumbling out. She told about the missed schedule and her call to the Sheriff. Finally, she related the dramatic story of her brief contact with him this morning. The reporter seemed intrigued as she explained about Morse code and how she had made the meter bounce out her message.

The reporter interrupted her a few times to ask questions about Amateur Radio and about the kind of equipment Marc must be using on the mountaintop.

121

"Why doesn't he just tell someone where he is?" she asked Kim.

"That's what we all wish he would do," Kim replied, "but his transmissions have been very faint and garbled. I think he must be so badly hurt or sick that he doesn't really know what he is doing."

Then she told about the Vienna, Austria bit and how Sergeant Willis had said he must be thinking about his parents. "I guess the cold can do strange things to your mind," Kim said.

"Isn't there a way to zero in on his signal like they do with downed aircraft?" the reporter asked.

"Well, yes there would be—but he would have to transmit more frequently than he is now for someone to get a fix on him. I know he's working on battery power too, so who knows how much transmitting time he has left," Kim sighed and visibly shuddered in front of the camera.

"Sergeant Willis has already told us about the search efforts underway. What do you plan to do now, Kim? Do you think you can talk to him again?"

"That's why I brought my two-meter radio," Kim said, showing her the small black hand-held transmitter. She switched it on to simplex and explained that if Marc were transmitting within a few miles on the same frequency, she would be able to hear him.

"And if you don't hear him?" the reporter asked.

"I'm not sure—depends on what the searchers report, but I may go back home later to listen on 80 meters some more."

Then the reporter asked one last question which caught Kim off guard and made her eyes fill with tears.

"You've never met this young man in person, Kim, but it's obvious that you're very concerned about him. Do you have any idea what he's like?"

"Special," Kim said, "and tough, I hope," she added in a whisper.

Chapter 17

SEARCHING

Thursday, May 18th
5:00 AM PDST
0500 Military Time
1200 Zulu Time

On Horseback
"Easy there, Jake, old boy—just back up easy."
Bill Harris edged the big black-and-white Appaloosa gelding down the horse trailer ramp. The horse obediently backed down the ramp and then turned to nuzzle his owner.

"Sorry I'm late," Bill yelled to his companions who were already saddled up and ready to go. The call for the Mounted Posse had come late yesterday afternoon. Unfortunately, he had been out of town on business, and his flight back from Phoenix didn't reach Portland until 10:00 PM. His wife had given him the message when he got home.

He quickly called several of the other men he worked with on searches. Apparently, some of them had been unavailable, too, but it was hoped there would be enough for a full search by morning. Bill was being asked to report directly to the Base Camp near Pamelia Lake.

"Not much sleep for either of us, huh, Jake?" he said as he slipped the hackamore over the willing horse's head. At over sixteen hands tall, Jake was the biggest horse in the Posse. *And the best*, Bill thought to himself. The sixteen-year-old gelding was as dependable as a horse could be. Everyone was glad when Jake was on a search because he seemed to exert a calming influence over the other horses.

Bill looked at the mounded snow on the ground and the threatening sky and was glad of his sturdy horse. If the going really got rough, Jake was definitely the horse to be on. Bill thought briefly about a young mare he had once ridden on a

search. The short horse had actually gotten stuck going over a fallen log and for a few minutes had teeter-tottered back and forth with her feet above ground on both sides. The trail had been narrow, and the horse's seesawing motions had made her come precariously close to slipping off the log into the canyon below. Bill and another rider had worked feverishly with ropes to pull her back to safety. She was rescued, but her belly was so scraped up she couldn't continue the search.

No such worries with Jake. If Jake couldn't clear an obstacle, no horse could.

Within ten minutes of his arrival, Bill and the two other riders were on the trail, following Sergeant Willis' instructions to search the area to Pamelia and then to radio back for further instructions. The Explorers would be following them for a more detailed inspection of the area.

Bill settled into the well-used leather saddle as they plodded up the slight incline. He would be watching the left side of the trail for the first three miles, hoping to see some clue of the lost young man.

His stomach growled slightly. A cup of coffee and a doughnut consumed on the way up had been his only breakfast. His saddle bags were packed with provisions—they, along with his sleeping bag and other camping gear, were always kept ready in the cab of his truck.

Bill's briefing on this case had been just that—brief, but the other riders filled him in on the details. They were looking for a 19-year-old college student who had been out here presumably since Saturday. Even though the search had just been started, Bill had sensed a note of urgency in everyone's voice. There had been reports that the lost student was injured, and in this kind of weather, that meant serious trouble.

Bill pulled up the back of his collar to cover his neck, too late to escape a blob of snow that the wind jolted out of a tree branch. He wiggled as he felt the cold crystals slip down the inside of his flannel shirt. Jake turned his head to look at his master.

"Boy, you don't miss a thing do you old fella? If I even twitch, you want to know why." Bill patted him on the neck, and Jake, satisfied that everything was all right, turned his head back to the trail.

They reached the lake without spotting anything out of the ordinary. Bill radioed back to the Base Camp while he studied a map of the area. The instructions came back to proceed up the trail behind Pamelia toward the Crest Trail and then to radio back once they had reached that objective.

On the backside of the lake, they came to a brief obstacle when the lead horse, a sorrel quarter horse, refused to go over a narrow log bridge spanning a rushing creek.

"Let's see if Jake will do it," Bill suggested. He got off and led his horse across the slippery log. Jake hesitated in the middle of the log and then gave a big leap to the other side.

"I'm not sure I'd try it," Bill said to the others. "The slope looks gentle enough—why don't you just take them across the creek."

The two other riders climbed back on their horses and made their way tentatively down the snowy creek embankment. The horses didn't seem too happy about sloshing through the icy water, but soon they were across and scrambling up the other side.

Snow had begun to fall. It was hard not to admire the beauty of the swirling flakes falling against the backdrop of tree-blanketed mountainsides. But the beauty had a sinister side, Bill knew—a side that could kill someone stranded in its icy loveliness.

"Come on Jake," he said, urging his mount up the steepening trail that led from the lake to the ridge above.

* * * * * * * * * * * * * * * * * * *

On Foot

The cold, damp air stung sixteen-year-old Julie Myer's face as she started up the trail toward Pamelia Lake. It was daybreak, but the only evidence of the sun's presence was a slightly lighter hue to the clouds. She had been up for hours. In fact, she had hardly slept at all.

As a member of the Explorers, she was used to being sent on search missions. Still, every time the call came, there was a rush of adrenalin that put her in high gear. Wednesday night, her commander had phoned, saying that they would be leaving at 3:00 AM for a search near Pamelia Lake.

She had scurried around, finishing her homework. Her friend, Megan, who lived next door, agreed to turn it in for her at school. Next, she wrote a note to her teachers. She had standing permission from them to be absent when needed for searches. Whenever she was gone, she had to work extra hard to make up her studies, but she was a good student and managed to keep a 3.5 average.

Boy, her pack felt heavy. She had been told that they might be spending one or two nights on the trail. That meant bringing a sleeping bag plus provisions and extra heavy clothing. If they were in for an extra long search, the Horse Posse would pack in anything else they needed.

Tufts of cottony snow adorned the unfurling spring ferns alongside the trail, and blue jays broke the early morning silence. Julie ignored the beauty of the quiet morning and concentrated her attention on the task at hand. Over 255 hours of training had taught her to be an efficient searcher.

When she had first joined the Explorer group at age 14, she knew nothing about searching. Her enjoyment of the outdoors and a desire to be of public service had interested her in the group. After the first meeting, she was hooked.

It was a long first year in which she had learned first aid, mountain climbing (both rock and snow), radio communications, map and compass reading, leadership, and survival training— culminating in a four-day solo stay in the wilderness where she built her own shelter and was allowed to take only six ounces of food. Over half the rookies had dropped out. Julie had thrived on the challenges, and now she was proud to be a Senior Explorer.

Twelve of them had arrived at 5:00 AM and had divided into three groups. Sergeant Willis had assigned them the trails to be searched, and they had set off shortly after the Horse Posse. The Explorers would be moving much more slowly and would search both sides of the trail thoroughly. In her group, Julie was assigned the left side of the trail. An Explorer advisor, a college student named Joe Brookings and a senior named Jill Smitke were on the trail itself. Another Senior Explorer, Tim Hauck, was on the right of the trail.

Their search was going to be difficult. The young man who was lost had walked in on a dry trail, and now snow covered any footprints he might have made. Still, there should be signs— perhaps some broken twigs, a food wrapper, remnants of a campfire—maybe even footprints around a streambed if they could find a muddy area the snow hadn't covered.

They moved slowly. Julie carefully climbed sideways down the embankments to explore the area 100 feet from the trail. They were instructed not to climb down any slope of more than twenty degrees. It was unlikely that a hiker would have strayed from the trail where the footing was treacherous, anyway. Still, Julie kept her eyes alert for any signs—even on the steep parts.

They walked in silence. The trail to Pamelia was gentle with many side trails and open areas where a person might have stopped to rest. At one place near a large grouping of rocks, Julie found a cigarette wrapper.

A quick call on their portable radios back to Base Camp. Did the lost man "Marc Lawrence" smoke? There was a pause as the React Communications person went to ask Marc's roommate.

The answer came quickly. No, Marc had never smoked in his life.

"Well, I guess someone else has been along this trail recently," said Julie to her companions. "That means anything we find may have been caused by either that person or by Marc."

They continued on. Such news was certainly nothing new. In the summer, they often were sent out over well-traveled trails. It was amazing that they were able to isolate clues from the person they were looking for. But they did, and their keen eyes often resulted in many rescues.

The lush spring growth had cascaded across the trail. Julie could see the neat hoofprints of the Horse Posse that had preceded the Explorers, and in many places, branches had been pushed aside by the moving horses. Since the Horse Posse riders were the first ones through, they would have noted any obvious tracks or signs. Apparently, there had been none, or Base Camp would have notified the Explorers when they checked in.

No, the signs were going to be much less obvious. Julie paused a moment, looking down a gentle embankment that led to the stream. The beauty of the setting was overwhelming. Ferns and rhododendrons flanked the rushing water. Snow covered the boulders that bordered the water—not a very inviting place to sit today, but what would it have been on the day Marc walked in? She closed her eyes and imagined sunshine streaming down through the tall firs and reflecting off the water. Would she have stopped there a moment to enjoy the scene? Yes. Would Marc? She crunched slowly through the crisp snow to check out the area.

* *

By Air

"Clear for takeoff."

The magical words had come just seconds before. Chief Warrant Officer Carl Blankenship, N7BAX, lifted the 8500 pound UH-1 Iroquois "Huey" off the ground at McNary Field and proceeded east. With him were Chief Warrant Officer John Whitten, co-pilot, Crew Chief Sam Billings, and Lieutenant Baxter from the Sheriff's Department.

The National Guard had received word early in the morning that an air search was wanted, but it was 1100 hours (11:00 AM) before the weather had cleared enough over the mountains to make it possible.

Pilot Carl Blankenship scanned the horizon as his craft zipped along toward the search area. The weather ceiling was low, and it was possible conditions could worsen again rapidly. They were hoping to make an hour search, but luck would have to be with them to have that much time.

Lieutenant Baxter talked to the pilot through the headset in his helmet.

"He's a ham operator, like you, Carl. Been out there since Saturday. All of our reports so far have been from hams who have heard him."

"What frequency is he on?" Carl asked, keenly interested.

"Code on 80 meters, but his transmissions have been very sporadic. No reports at all today since early this morning."

Carl nodded. It was difficult to really talk in the noisy helicopter, but he mulled over what the lieutenant had said. Too bad the lost young man had gone off the air. Otherwise, they could home in on his signal with their ADF (automatic direction finder). They would need a fairly frequent signal for them to attempt any kind of zeroing in.

"Okay, this is the area," Lieutenant Baxter announced from the rear seat.

Carl slowed the helicopter dramatically and dropped down to where they were almost skimming the treetops. The four men peered out the wide windows, looking for the slightest sign of human life as the craft began a slow north-south and then east-west search of the grid pattern Lieutenant Baxter had set out on his map.

Co-pilot John Whitten was a Vietnam veteran and had participated in countless search and rescue missions there. If anyone could spot a person from the air, Carl knew John would be the one. He urged the helicopter down into the gullies and draws, hoping to see anything—the remnants of a campfire, Marc's bright orange jacket, a sign of any kind.

It would be wonderful if they could find something obvious— like the lost hunters who had written "help" in bright orange Tang in the snow. That was easy. Carl somehow knew that this search was going to be anything but easy.

They were only half done with their methodical grid search when it started to snow again, and fog closed down their visibility considerably. Lieutenant Baxter was talking on his hand-held radio to the Base Camp. He had an earpiece attached to the speaker, but even so, it was difficult for him to hear over the noise of the rotors—probably even more difficult for them to hear him.

Lieutenant Baxter reached forward and tapped Carl on the shoulder.

"They just sent a new weather report into the camp—this stuff is supposed to get thicker for the rest of the day. Guess we had better go back while we still can."

Carl nodded and reversed his course. They climbed a few hundred feet, bouncing through the rough storm air. It was imperative not to get stuck in the storm. He breathed a small sigh of relief as they neared the edge of the wilderness area, and the cloud ceiling began to rise. It always amazed him how the weather could be so dramatically worse just fifty miles from Salem.

Probably surprised our lost young man, too, he thought shaking his head.

What was his call? KA7ITR. That was it. Well, KA7ITR—let's hope for a contact real soon, he said to himself.

They were soon on the ground again in Salem.

"Stay on alert, Carl," Lieutenant Baxter told him. "If the weather clears, we'll want to go back and finish that grid."

Chapter 18

IN A FARAWAY FOREST

Thursday, May 18th
12 noon PDST
1900 Zulu

But the weather didn't clear. The moving front that had forced the National Guard helicopter back to base thickened and swirled and dumped its snowy load on the waiting treetops of the mountainous forests below. Marine air from the Pacific Ocean combined with cold Arctic air to swell together in an unseasonable storm that covered virtually the entire Northwest. Nature, as if apologizing for the lack of moisture during the winter and spring, blew rain into the dry valleys and snow into the mountains.

Deer pawed at the thick white powder to get at the young spring grass underneath. Rabbits, foxes, skunks, and an occasional porcupine scampered playfully through the white carpet. But one animal underneath the trees was not moving at all: a human named Marc Lawrence.

The trees had protected Marc from direct snowfall, but blowing gusts had almost completely covered the lower half of his sleeping bag and mounded the snow on either side of him. Just before dawn, in what had been his last truly lucid moments, he had somehow been able to clear his foggy mind enough to transmit to Kim. It had been a terrible effort to drag both the HW-9, with its antenna tuner and flashlight power supply, and his two-meter rig up onto his chest, but somehow he'd managed to do it. He thought he'd seen a response from her, but he wasn't sure. His vision was very blurry. The clearing in his mind had been like the clearing in the storm—brief.

Realizing that he was fading fast, he had turned on the two-meter rig which he had set the other day on the simplex frequency 146.52. Weakly, he'd croaked a "help" into the microphone. No answer. He turned it off, afraid that he would go back to sleep and exhaust the battery pack.

He lay there, willing his hand to send more code, but nothing happened. It was as if the relays between his brain and hand were frozen. At daybreak, he drifted off into a semi-slumber. Occasionally, he opened his eyes and stared blankly at his surroundings, but he made no attempt to move or think. He was merely a dulled spectator, dimly watching a snowy landscape that seemed disconnected from him.

As his movements lessened, the snow crept in. His body and his mind were quieting into what could soon become a fatal, cold sleep.

For the past two days, his infected, tortured body had alternated between spiking a fever over 104 degrees and intense shivering. The combination had completely exhausted him. Not only was he at risk for hypothermia, but dehydration as well. The effort of his body to keep its temperature in a normal range near 98°F had taken its toll. He had put up a valiant fight, but now he was losing. And losing rapidly.

As his movements slowed, the cold snow penetrated the sleeping bag, and his body gradually cooled. Hour by hour, his temperature dropped until it was close to ninety-one degrees Fahrenheit. His heart and breathing rates slowed and, with them, his thought processes. He was no longer aware of where he was or what he was doing.

As Marc slipped into a coma, his Amateur Radio gear rested on his barely moving chest. One quick turn in the sleeping bag, and it would all have tumbled to the side. But, Marc wasn't turning. Except for his hands which occasionally moved restlessly, he was still.

He was drifting and it felt so good. He didn't have to fight these terrible cold and pain monsters anymore. Everything was going to be okay. Peace was his as his body temperature continued to creep downward. His temperature was 90.5°F.

But somewhere in his mind, the Marc that everyone admired for his strength and stubbornness, refused to let go. Whatever that part was, it held on to the delicate life thread

that was still his. Marc dreamed. He was hiking with his father. They were in a meadow. The sun was shining. The warmth felt good. Marc could see mountains in the distance. Everything looked familiar. Of course. They had been here before. They were in Hunt's Cove.

The images dropped from his mind, but the words Hunt's Cove remained. He must tell someone those words, but he didn't know why. •••• *dit dit dit dit (H)* ••— *dit dit dah (U)* —• *dah dit (N)* — *dah (T)* ••• *dit dit dit (S)* —•—• *dah dit dah dit (C)* ———— *dah dah dah (O)* •••— *dit dit dit dah (V)* • *dit (E)*. The letters moved slowly through his mind, and almost as if by reflex, his hand on the key moved in rhythm with his thoughts. The all-important message was sent just once.

Marc's body temperature reached 90 degrees and hovered there as blood moved from his skin to the core of his body in a valiant attempt to keep his vital organs still functioning.

His right hand rested on his chest, his increasingly white fingers still loosely clasped on the key. The snow swirled. Blue jays squawked as they flew to and fro. The world listened anxiously for a further message as Marc drifted into a white night.

* * * * * * * * * * * * * * * * * * * *

2100 hours Austrian time (9:00 PM)

"Did you get him?" Mrs. Lawrence asked her husband as he came back to the table. They were getting ready to have dinner in a small lakeside cafe in Millstatt am See, a delightful old village in Austria's southern province of Carinthia.

The Lawrences had been so entranced with Austria that they had cancelled a couple of their side trips in order to explore the varied countryside more thoroughly. Today had been particularly memorable. After touring the 1000-year-old buildings and admiring the fifteenth-century fresco that was the town's pride, they had spent the rest of the day walking in the woods. The weather had been perfect—65 degrees and clear. The 35 miles of well marked paths through the lakeside woods were all beautiful and made choosing a trail difficult. The Lawrences had walked until their feet tired.

After a brief nap in their hotel room, they had come here for dinner. First course, a rich "Nudelsuppe" (noodle soup) had been served. It was after the waiter had carried away their empty bowls that Mrs. Lawrence had suggested that her husband try to phone Marc.

"I know—we were just going to write him postcards, but I dreamed about him last night, and I would just feel better if we checked with him. It's almost time for his final exams and he might appreciate a pep talk."

Mr. Lawrence had willingly gone to call his son. Funny thing. Marc had been in his thoughts all day too. As much fun as this trip was, it would be good to get home and see him again. Summer was coming, and he hoped they would have time for some fishing and camping trips together.

"There was no answer," he told his wife as he sat back down. The waiter had just arrived with fragrant Wienerschnitzel—a dish that was a favorite of both of them.

His wife glanced at her watch. "Ought to be about lunchtime in Oregon," she said. "Since it's Thursday, I thought we might catch either Marc or Mike in their room—I know that's when they usually do their lab assignments together."

"Well, they're probably at lunch. I'll try again in another hour," Mr. Lawrence said.

They ate slowly, savoring the rich dish. The view of the moonlit lake was magnificent, and Mr. Lawrence reached over to hold his wife's hand.

"A perfect day," Mrs. Lawrence said, sighing. "Oh no," she groaned as the waiter came carrying a dessert tray.

"How can we possibly eat anymore?" she said. But they did. They each selected their favorite—"Indianer"—a round chocolate-covered puff that was about fifty percent whipped cream. The waiter kept their cups filled with fragrant coffee.

A small band was warming up in an adjoining room, and the lilting refrain of a Viennese waltz filled the cafe.

Mr. Lawrence reached for his wife's hand again.

"Would you care to dance, my dear?" he asked, trying to imitate an Austrian accent.

"My pleasure," she said, giggling at his formality. He put his arm around her and together they walked onto the dance floor.

* * * * * * * * * * * * * * * * * * *

Pamelia Lake
12:30 PM PDST

"I feel like a fifth wheel," Kim confessed to Mike as they stood outside the communications van at the Base Camp. "Wish there were something I could do to help."

"It looks like they've got everything under control," Mike said.

The tremendous flurry of activity that had been at the camp earlier had now settled down into relative calmness. All of the patrols were out on the trails. The helicopter had made a search and turned back due to weather. There was nothing to do now but wait for some word from a search team or a break in the weather so that the aerial search could resume.

Mary came to the door of the trailer and talked to them.

"You're welcome to come back in here and wait where it's warm," she said. "I doubt too much is going to happen until this storm front passes through—it makes really slow going for the Explorers and the Posse when the snow is blowing like this."

"I know; thanks, Mary. We were just stretching our legs," Kim said. She watched the snow blowing down sideways, and reached up to wipe the flakes from her face.

Just then the crackling of the radio caught Mary's attention, and she rushed back inside to answer whoever was checking in. Kim and Mike followed her and sat down just in time to hear Lieutenant Baxter's voice on the air.

"Base Camp, we've just had a call from an Amateur Radio operator in the ARES Net in Marion County. He's been monitoring for Kim—didn't catch his call. He said he heard a very weak transmission just a few minutes ago that said "Hunt's Cove"— wonders if it could be our boy?"

Thank you, Pete, Kim said silently to herself.

"Go get Sergeant Willis," Mary told Mike as she began poring over the map of the area.

Sergeant Willis came running back with Mike. He didn't need to look at a map.

"Hunt's Cove—that could be him all right. The cove area is about seven miles from here, up near the Crest Trail."

His normally relaxed tone became crisp and business-like as he gave orders to Mary.

"First ask Lieutenant Baxter to tell the ham operator to keep monitoring—if we can get any more information, it will certainly help." Sergeant Willis scribbled some notes on a pad next to the Jefferson Wilderness map. "When Explorer Group One checks in, advise them to mark their position and then proceed to Hunt's Cove to check out that area. Posse Three should come in from the south and meet them there. Have them do a hasty search of the area and then report back."

Kim and Mike waited in the small trailer for the next hour while groups checked in and Mary relayed Sergeant Willis's orders.

Finally, when Kim felt she couldn't stand the tension of waiting any longer, she and Mike went outside to stand under the trees for a few minutes. The snow was coming down heavier than ever.

Sergeant Willis, running back to the communications van after talking to a Jeep Patrol member who had just driven into camp, shouted to them.

"Sure wish this weather would clear. This is getting worse by the minute."

Mike and Kim exchanged worried glances and, without speaking, they walked over to the kitchen trailer to get something hot to drink.

Chapter 19

TOUGH DECISIONS

Thursday, May 18th
2:00 PM PDST
1400 Military Time

Lieutenant Baxter was no stranger to tough decisions. His ability to think and act well under pressure was what had earned him several military awards in Vietnam. His men had always depended on him, and he'd never let them down.

His role as Search and Rescue Coordinator for the Sheriff's Department was also one which required that he make tough decisions. Lives depended on those decisions. Not just the lives of victims, but the safety and well being of the search units he put in the field. On Thursday May 18th, the weather was forcing him to make choices he wished he didn't have to make.

Weather reports had been coming in all morning. He didn't need a report to know that they were in the middle of an intense storm system. Through his office window, he could see unrelenting rain pelting down on the streets. It was coming down so hard at times that motorists were pulling over because their windshield wipers couldn't keep up with the deluge. Lieutenant Baxter could just imagine what all that moisture translated into snow in the mountains would mean.

He called the weather bureau out at the air field, hoping to get some encouraging news about the storm letting up. Yes, he was told, the storm would let up—but not until tomorrow morning. For tonight, it would be windy and rainy in the valley with lows down to the thirties. In the mountains, there would be heavy snow with temperatures down to 20°F.

He looked at the clock on the wall. It read 2:00 PM which his military training promptly translated into 1400 hours.

There was still time for the Explorers and the horses to check out Hunt's Cove to see if possibly their victim was there. Maybe they would get lucky and that would be where Marc Lawrence would be found. But if he wasn't there, Lieutenant Baxter would have to call his searchers back to Base Camp for the night. They had too little equipment with them to camp in the snow.

* * * * * * * * * * * * * * * * * * * *

Posse #1 reached Hunt's Cove about the same time that Explorer Team #2 did. Bill Harris and his group had reached the Crest Trail and were heading North away from the Hunt's Cove area. Bill had stopped the group in the snowy shadow of Mt. Jefferson to radio back to Base Camp. The orders that came back were brief and to the point. Mark their spot and then head directly to Hunt's.

It was slow going along the ridge. The Posse reversed direction and fought their way back toward the first trail that led to Hunt's Cove. The snow was deep, and the wind was blowing so hard that it was difficult to see. Bill led the group with Jake, who put his head down against the wind. The others followed him obediently. Bill and the other riders encouraged their horses, speaking to them and patting them on their necks frequently. As they struggled through the drifts, the riders had no way of knowing that they were within two miles of Marc.

In fair weather, had there been any reason to believe that the hiker had taken a shortcut from the Crest Trail to Hunt's Cove, the searchers also would have left the trail and might have come across him. In a near blizzard, they had no choice but to stick to a path that was somewhat visible. "Visible" was a questionable word. Even though the riders were familiar with the area, today it was hard to see which way the trails led at all. They depended heavily upon their horses to pick the path. Jake led the entire way, picking his footing with the care of a mountain goat.

140

It was a roundabout trek that finally brought them to Hunt's Cove. Explorer Team #2 had actually been headed up the switchback trail from Pamelia to Hunt's when they made their required radio check-in with Base Camp and found out their new orders.

Julie Meyers of Team #2 felt more than just a little apprehensive as she trudged up the last 100 yards to Hunt's Cove. Her legs ached, and her lungs felt as if she had run a marathon. Climbing the steep, slippery trail had taxed her stamina. As a member of her high school track team, she ran daily. But nothing could totally prepare her for the rigors of a hike like this with the extra weight of her pack and equipment.

The weather seemed to be getting worse every minute. She certainly hoped they wouldn't have to spend the night out here, but if they did, the Posse would bring them gear.

Overriding all of her own physical fatigue was excitement and hope. It was possible that they would find the lost man at this location. Then all their efforts would be worthwhile. Rescue was their driving purpose, and she focused on it when the going was especially difficult.

The Explorers crunched through the snow into the meadow and laid down their backpacks under a snow-laden tree. Bill and the other Posse members were already quickly scouting the perimeter of the meadow. Every so often the Explorer team leader would call out "Marc on three!" Then, all four Explorers would shout out "One, two, three, Marc!" into the snow-driven wind. There was no answer.

Julie and the others circled the meadow, widening the search by also going into the woods. For more than an hour, they explored the area thoroughly. The best they found was a depression in the snow near the trees. They dug down and found the remnants of a campfire. Was it Marc's? It was impossible to tell. They radioed that information in, but Sergeant Willis didn't seem to think it was significant. Unless the fire was less than a day old, it would be impossible to tell whether it had been from last Saturday or last September.

The three Posse members and four Explorers huddled together under the tree where they had dropped their backpacks.

"There are lots of directions he could have gone from here," Bill Harris shouted from aboard Jake. "I just have a feeling he's close, but we'll have to get orders from Base Camp before we can go any farther."

The orders that came from camp were disappointing. "Return at once." The weather was predicted to be worse through the night. Everyone was to spend the night in Camp and hopefully resume the search at daybreak.

It was a hard decision to accept. The seven of them stood silently. They knew the decision had been made with their safety in mind, and they also knew how to obey orders. In unspoken agreement, they made one final sweep of the edge of the meadow in case they had overlooked something. The sound of "One, two, three, Marc!" echoed through the valley as they slowly made their way back to the trail.

The Posse left first as they were to return by a different trail that was yet to be explored. The Explorers took the shortest trail, but even that was over seven miles. For a brief moment, Julie wondered if she had the strength to make it back to camp.

She pulled her wool stocking cap down on her forehead almost to her eyebrows. She shouldered her pack and tried not to think about her aching legs as she set off down the trail.

* * * * * * * * * * * * * * * * * * *

Kim and Mike's faces clearly showed the disappointment they felt when they heard Sergeant Willis's order for everyone to return to Base Camp.

They listened to a few of the units checking in and then they went outside to talk. Sergeant Willis came out the door of the communications van and came over to join them under the trees where they were somewhat sheltered.

"You can't just give up and leave him out there," Kim said accusingly. "He's going to die if you don't get him quickly."

Sergeant Willis looked at her sympathetically. He didn't seem to be angry at her outburst, but he was very firm when he spoke.

"Kim, right now, I have 22 people I'm concerned about. Nine of them are on horses, twelve of them are on foot, and one of them, Marc, is on the ground. They're all equally valuable to me, and I can't risk one for the other. I understand how you're feeling. I want to get him out of there, too, but we're not going to risk losing someone else to do that."

Kim didn't say anything as Sergeant Willis walked back to the communications van for more check-ins with units.

"I wish we could go look for him ourselves," Mike said.

Kim nodded and looked across the camp area at the trail leading toward Pamelia Lake. The snow was blowing so hard that she couldn't see even past the first tree.

"You know that's impossible and so do I," she said to Mike sadly. "I'm sure the Sergeant's right, but I wish he weren't."

A four-wheel drive truck entering the camp caught their attention.

"Oh, it's Sharon," Kim said, recognizing her ham friend from the Jeep Patrol. She watched as Sharon disappeared into the communications van. In a few minutes, she came back out, and Kim walked over to talk to her. Sharon greeted her with a hug.

"I thought you might be here, Kim. I tried calling you on two meters a couple of times as I was driving up from Salem this morning, but I didn't hear you."

"No, we didn't get here until later. I suppose you know everything's shutting down for the night."

"Yeah, Sergeant Willis just told me. In fact, I'm on my way back to town—there doesn't seem to be any more for the Jeep Patrol to do tonight. We've gone over every road there is several times."

Kim shook her head sadly. Sharon looked at her with motherly concern.

"What about you, Kim? Are you going to spend the night here, or what?" Sharon wanted to suggest that Kim go home for the night, but she could tell by looking at Kim's face that she needed to make those decisions herself.

"I don't know, Sharon. If only I had an 80-meter rig here, I would stay. I want to listen for Marc, and yet I want to stay here, too."

"Kim, I have a portable rig. Why don't I go get it for you and bring it back? I'm sure there's room to set it up in the van, and we can string an antenna."

Kim's face lit up.

"Would you? That would be great. In fact, could I go with you and pick up my sleeping bag and some more clothes at my house?"

In her excitement, she had completely forgotten about Mike who was standing there. She turned to him apologetically.

"What do you want to do, Mike? I don't want to keep dragging you back and forth just to get my stuff."

"No problem, Kim, but I tell you—if Sharon can take you down and back, I think I'll go back to school for the night and come back in the morning. Is there any way you can call me if something happens?"

Sharon answered the question. "We can get a message to the Sheriff's Office, and they would relay it to you. If you need to get a message to us, do it the same way in reverse. Just call the Sheriff's Office and ask them to send the message to Base Camp." Kim talked to Mike alone for a minute before he left.

"Thank you, Mike—I appreciate everything."

"Hey, you know, he's my friend, too. I just wish there were something we could really do to help."

"Maybe there will be, tomorrow," Kim said. "Maybe tomorrow."

* * * * * * * * * * * * * * * * * * * *

Kim's parents were just getting home from work when Sharon and Kim drove into the driveway. They had stopped at Sharon's and loaded the 80-meter transceiver, antenna tuner, and antenna wire into the back of her truck. Kim introduced Sharon to her parents.

After a quick review of what had happened during the day, Kim told them her plans about going back to the camp. To her surprise, her mother and father looked disapproving. She hadn't expected this, as they had been supportive all along.

"Hey you guys, what's wrong? You know I have to do this!"

"Kim, go look in the mirror and you'll see what we see—one very tired young woman. And look at the weather! To say nothing about the fact that you have school tomorrow."

Kim's mother looked at Sharon for support. Sharon just shook her head. On the way down, she had heard Kim's determination loud and clear—there was no way she was going to enter into this argument. What she could do, though, was reassure Kim's parents.

"If you do decide to let her go back up, I promise you she'll be well looked after. There's hot food in the camp, and Sergeant Willis looks after everyone like a parent. Kim showed me her sleeping bag, and I'll be happy to let her borrow one of mine which is heavier."

Kim gave Sharon a grateful glance. Finally, Mr. Stafford spoke.

"Go ahead, Kim. And good luck."

Kim ran to get her things.

* * * * * * * * * * * * * * * * * *

High in the mountains under a row of trees near a frozen stream, there were no decisions left for Marc Lawrence to make. He lay unconscious, drifting farther and farther away from any hope of survival.

Chapter 20

NIGHT WATCH

Thursday, May 18th
8:00 PM PDST
2000 Hours Military Time

"Τ**he search continues for 19-year-old Marc Lawrence,
an Oregon State University freshman and Amateur
Ham Radio operator who has been missing in the
Jefferson Wilderness area since Sunday. Amateur Radio
operators who have heard his intermittent distress signal
believe that he is injured and suffering from hypothermia.
While searchers comb the woods, ham radio operators all
along the West Coast are monitoring the frequency where he
was last heard. A local ham, 17-year-old Kim Stafford of
Salem, alerted search agencies to the possibility of his being
lost after he did not show up for a scheduled conversation on
the air last Sunday. The National Guard is hoping for a break
in the weather early tomorrow so that they can resume their
aerial search."

A commercial came on the radio station, and Sharon
reached over and turned it off. Kim sat silently in the truck
as they made their way back up to the camp. She had picked
up the morning newspaper at home and had been surprised
to see her name mentioned in the story about Marc on the
front page. The headline read "Local 'Ham' Helps in Rescue
Effort."

"That should be 'hams', not 'ham'," said Kim. "Look at all
the people who are monitoring right now and the ones that
have called in reporting his signal before."

"You really don't like being in the limelight, do you?"
asked Sharon sympathetically as she reached over and patted
Kim's hand.

"No, I don't," Kim said emphatically, "but all that matters is saving Marc. His rescue is the only story I'm interested in reading."

They were approaching the Detroit Lake area, and Sharon slowed down and put the truck in four-wheel drive. The road had been plowed, but the steady snowfall was keeping it slippery.

Kim's stomach growled so loudly in the silence of the truck cab that they both laughed. Sharon glanced at her watch.

"7:30—when did you last eat, Kim?"

"I had a sandwich at noon."

"Well, we'll fix that as soon as we get to camp. Bernie's a super cook, and I imagine he has something tasty cooking."

They bounced along the access road to the Base Camp. The lights of several campfires greeted them as they rounded the final curve and parked the truck at the perimeter of the clearing. The camp seemed twice as crowded as it had that morning. Ten horses, sporting colorful warm blankets, stood tethered under the trees, munching on a bale of hay that had been spread along the ground for them. A large campfire blazed in the middle of the grounds, and quite a few adults stood near it, drinking coffee and talking.

Kim and Sharon walked over to join them.

"Where's Sergeant Willis?" Sharon asked.

"On the foot patrol," someone laughed.

Kim looked at Sharon questioningly.

"Sergeant Willis looks after the Explorers like his own kids," Sharon explained to her. She pointed to the edge of the camp where many of the Explorers had set up mini camps with their own fires. "After a long day of hiking, their feet are often wet from perspiration. He makes sure that each of them washes and dries his feet and puts on dry socks."

"It's not a suggestion either—it's an order," someone else chimed in.

Apparently, everyone's feet had been inspected because Sergeant Willis came back to the campfire, smiling.

"You'd think they'd be tired," he said, shaking his head. "But they seem to be full of energy. I hope they—and we—get some sleep tonight."

He walked over to stand by Kim and Sharon.

"Speaking of sleep," he said, accusingly to Kim. "I see you came back. That must mean you're not planning on getting any."

"Sharon has loaned me her portable 80-meter transceiver so I can monitor for Marc," Kim said. "I just want your permission to set it up in the communications van."

"Are there others listening for him too?" Sergeant Willis asked.

"Oh sure, we put a bulletin out on the ARES net, and I think the Salem group has divided the night into shifts. Plus, I'm sure the word has spread to other areas where he's been heard."

"But you want to listen, too—just to make sure—right?"

Kim looked at him to see if he were teasing or possibly angry, but his face was serious.

"Right," she said. "And I wanted to be close to the action too. I just feel that I have to be here. I hope that's okay."

"I don't blame you, Kim. Let's get someone to help string your antenna and get you set up."

"She needs some dinner, too," Sharon said.

Everything was taken care of in record time. A React (communications volunteer group) member helped string the 63-foot antenna. Sharon and Kim carried in the gear and set it up in the communications van. Mary Hammond, the operator Kim had met, was sleeping in her truck. Tom Forbes was the night operator, and he greeted Kim with a warm handshake.

"Put all your stuff down at the end of the table, and I'll move our gear this way as much as I can," he volunteered.

Soon, the equipment was all in place. Since it was Sharon's rig, she sat down and tuned the transceiver up carefully on 80 meters. A few quick explanations and Kim was set to operate.

Bernie, the camp cook, appeared in the doorway with a steaming bowl of his famous stew. Kim gratefully accepted it and sat down at the rig. She put headphones on so as not to disturb Tom while he monitored his various frequencies.

Sharon tapped her on the shoulder.

"Everything okay?" she asked.

"Fine, Sharon. Thank you for everything. Are you going home?"

"I guess I will, Kim. Sergeant Willis says there's nothing for me to do until morning. I'll see you then. Try to get some sleep— okay?"

Kim nodded and turned her attention back to listening for Marc.

She ate her stew quickly while she scanned the band and then warmed her hands around a cup of hot tea as she concentrated on the area of the 80-meter band where Marc had always been heard. An hour or so passed with nothing.

Occasionally, Kim tuned up and down the band. DX was good tonight—she heard a Japanese station talking to someone in Texas and several Hawaiian hams talking to different locations in the states. The urge to join in the fun of long distance contacts was only faint in her mind tonight. She had work to do.

She tuned back to "Marc's frequency" and dialed back and forth through it several times just to make sure Sharon's transceiver was calibrated the same as her own.

The first beep caught her by surprise and she sat up straight with such a jolt that Tom turned to look at her questioningly. She took her headphones off and turned on the speaker so he could hear. What she heard was definitely a Morse code signal, but there was no pattern to it at all.

It had the same odd chirping tone that she remembered about Marc's signal from the other morning when she had heard his distress signal. It was in the same place, but whatever this beeping was didn't make any sense. First a dash, then about a minute later another dash and a couple of dots. Kim tried copying them down to see if they made any letters at all. They didn't. They were just random sounds, but Kim felt sure they belonged to Marc.

Tom raised his eyebrows and looked at her for an explanation.

"I don't know—sounds like somehow he's holding the key and is unaware of what he's doing," Kim said. "Sure doesn't sound good."

The word spread through the camp that Kim thought she was hearing Marc, and soon many people stuck their heads in the door to listen. Everyone was puzzled.

"Do you really think that's him?" Bernie asked.

"Yes, I do," said Kim. "The signal is in the right place, and it has a chirp that I heard before. I believe some of the other hams who have heard him have reported the chirp too. I'm just sure it's him."

The signal continued, averaging about two or three beeps per minute. Sometimes there were silences for several minutes, but then the beeping would start again.

Around midnight, Sergeant Willis made a faint attempt to make her get some sleep.

"Look, suppose I have someone else sit here and listen and promise to wake you up if there's any change?"

"Does that someone else know Morse code?" Kim asked.

"No, but you said he wasn't sending anything intelligible," replied Sergeant Willis.

"I keep hoping," Kim said, showing him her scratch pad of letters she was trying to form out of his random sending.

"There are other hams listening, too," Sergeant Willis said. "You said that, yourself."

Tom nodded in agreement. Just then a call came from the Sheriff's Office reporting three phone calls from amateurs who had heard the beeping.

Kim looked up at Sergeant Willis. His gray-blue eyes looked stern, but kind. She shook her head and turned back to the rig. He put his hand lightly on her shoulder and then left the van. A few minutes later, he reappeared carrying the sleeping bag Sharon had left.

"Up," he said, motioning for her to stand up. "You can stay, but I want you to slip into this sleeping bag in your chair. That way if you do go to sleep, you'll at least be warm."

Obediently, she stood up, took off her shoes, and stepped into the heavy down-filled bag. He zipped it up the side, and she sat back down in her chair, feeling slightly like a mummy. It was warm—she had to admit that. It felt good to stretch her toes and have the warm comfort of the sleeping bag around her.

She scooted her chair closer to the table and put the headphones back on. The beeping was continuing. It had sort of a strange rhythm all its own. Kim tried timing the sounds to see if she could predict the next one. There was no pattern, but instinctively she knew when the next one was coming. She felt linked to whatever force it was inside of Marc that was sending a signal.

Occasionally, she tried transmitting to Marc, sending slowly as she had done at home. There was no response, and she came to the conclusion that whatever his condition was, it prevented him from seeing her messages.

The night drifted on. Noise from outside died down as the Explorers discovered that their warm sleeping bags next to a campfire were more sleep inducing than they had thought. Between 1:00 AM and 4:00 AM, Sergeant Willis didn't come into the van, and Kim presumed that he was getting some sleep too.

She and Tom chatted. He told her that he was a computer programmer for the State of Oregon.

"Mary Hammond's husband works in my office. He told me about Mary's work with Search and Rescue. I became interested and started volunteering three years ago. It's fascinating work—nice to do something where you feel useful."

Kim nodded. She was having trouble staying awake, and she welcomed the conversation. Tom asked her questions about her future. She told him about Oregon State—about wanting to become a veterinarian or marine biologist. And she told him the details of how she had met Marc on the air.

"I know it sounds odd," she said. "but I feel like I've known him for a long time."

Around 5:00 AM, Sergeant Willis opened the door and brought them both steaming hot cups of coffee.

"Are you still hearing him?" he asked Kim.

She turned the speaker on and let the erratic beeps speak for themselves.

"Good news," Sergeant Willis said. "The weather is clearing. I've talked to the National Guard. They've got a chopper ready to go that has an operational automatic direction finder that will pick up 80 meters. They're standing by to take off at first light."

Kim felt her fatigue evaporate.

"Then today is rescue day!" she said hopefully.

"I think so!" said Sergeant Willis. "At least I hope so."

Chapter 21

FRIDAY

Friday, March 19th
6:00 AM PDST
0600 Hours Military Time

Even Kim was surprised at the energy she felt as the sun came up. Her sleepless night was just a memory, and she felt equal to any task—especially the rescue of Marc Lawrence.

Everyone else in the camp was obviously feeling the same nervous energy. The Base Camp was buzzing with activity. Horses were being saddled, daypacks loaded with provisions, and heavy boots and clothing donned as the welcome sunshine spurred everyone into action.

Kim had to shield her eyes with her hand as she came out the door of the communications van because the sunlight glinting off the snowdrifts was so bright. She had left the volume on the transceiver turned up loud, and Marc's erratic signal was clearly audible outside the van.

Mary Hammond had come back on duty and was busy talking to Lieutenant Baxter who was getting ready to leave the Sheriff's Office for the National Guard airfield. The Guard air crew were set to resume the aerial search as soon as some local fog at the airfield cleared.

"Looks good here!" Sergeant Willis said, walking over and pointing at the sky, "but that doesn't mean it's the same in Salem."

The low gray clouds of yesterday were gone and had been replaced by big fluffy white ones which were rapidly scudding across the open blue expanse. "Weather bureau said the clearing may hold until this afternoon when the next system moves in."

All three of the Posse search teams were saddled and ready to go. Sergeant Willis talked to their leaders and discussed which routes they should pursue after they returned to the sites they had been searching yesterday. Two of the Posses started off on the Pamelia Creek trail where they would then branch off in opposite directions at the lake.

"I think I'll hold that third team back for an hour," Sergeant Willis said. "Let's see if our helicopter boys find anything. We may want to send horses in a totally different direction fast."

* * * * * * * * * * * * * * * * * * * *

All three of the Explorer groups had left the camp with directions to check in promptly every half hour in case there were new instructions. Julie Meyer's team had left an hour before the others—before daybreak to return to the Hunt's Cove area and search the surrounding areas. They had just reported in.

"They've already reached Pamelia Lake and are starting up the switchback trail. They're sure making good time," Mary Hammond reported.

Sergeant Willis exchanged a quick glance with Mary. Everyone knew how disappointed that particular Explorer group had felt at being called back yesterday. No wonder they were hustling today. If there were a possibility of their being called back again due to weather, the group wanted to make sure they had plenty of time to give Hunt's Cove a better search.

Kim stood in the van doorway and watched the departing groups wistfully. She had a real desire to go with them, but she didn't know if Sergeant Willis would let her. And which group would she ask to go with? Supposing she chose the one that went in the complete opposite direction of where Marc was? And what if Marc transmitted something of vital importance while she was gone?

No, she needed to stay here. But right now, she felt as if she weren't doing anything useful. She sighed and went inside to the rig. She sat down, picked up a pencil, and once again tried making sense out of the jumble of dots and dashes Marc was sending.

* * * * * * * * * * * * * * * * * * *

"I'll be there in ten minutes." Lieutenant Baxter put the phone receiver down and hastily grabbed his heavy field jacket. He signed himself out for the morning and left a note with the dispatch operator.

"Clearing—visibility improving." These were the magical words he had just heard from the National Guard Air Station. The Huey with the automatic direction finder was being readied for takeoff. Lieutenant Baxter took the steps outside the building two at a time and ran to his car.

So much moisture. He couldn't remember a spring storm that had waterlogged the valley the way this one had. Even though it had stopped raining for the moment, the water dripping off trees splashed on his windshield as he drove along State Street. Passing cars hitting puddles completely drenched his car, so he turned the wipers on high in order to see.

The airfield was only a few miles away, and soon he turned in the gate. The security guard waved him on through and he parked alongside the Administration Building.

Chief Warrant Officer Blankenship and his crew were already in the helicopter going through check-out procedures when he arrived. He ran across the taxi strip and ducked his head under the gust of the rotors.

Crew Chief Sam Billings gave him a welcome hand up through the doorway. Pilot Carl Blankenship, N7BAX turned around in his seat.

"Ready?" he said to Lieutenant Baxter.

"Ready. Let's find our man."

The helicopter's rotor speed increased as the crew awaited takeoff instructions from the tower.

* * * * * * * * * * * * * * * * * * *

Something was making noise inside Marc's head. Everything had been so quiet as his body temperature lowered and he drifted into oblivion. He wasn't conscious of his release from pain. His senses had just slowly dulled and then there had been nothing. It wouldn't be long until the cold penetrating through the core of his body cooled his heart and other vital organs down to the place where they stopped. For most victims of hypothermia, there would have been no return. But something was trying to draw Marc back. In the deep recesses of his subconscious, he was dimly aware of pain. And noise. There was a loud whirring sound somewhere above him.

Wake up Marc! Someone was talking to him. The shadowy image of his grandfather filled his mind, and Marc reached out his hand as if to touch him. The hand fell limply back on the code key, sending a long dash before he let up pressure on the key.

He was swimming through something white—trying to get to the surface even though it hurt. When he stopped trying, he started floating downward again, and the pain went away.

His mind fluttered toward awareness again. Something was forcing him to surface. And that something had a voice.

Send, Marc! Send 146.52—come on Marc •———— dit dah dah dah dah (one) ••••— dit dit dit dit dah (four) —•••• dah dit dit dit dit (six) •••• dit dit dit dit dit (five) ••——— dit dit dah dah dah (two). Come on Marc! Send it.

It was his grandfather, and he was insistent. Marc struggled with his grandfather's image, trying to make it go away. It hurt to be awake. He felt as if his grandfather were physically shaking him.

Do it, Marc! 146.52.

Why Grandpa? Marc formed the words silently in his hazy consciousness.

The reply was loud in his head. *Simplex, Marc. Talk to them. Tell them where you are. Live, Marc!*

Marc groaned and his eyes flickered briefly. He moved his hand slightly and grabbed hold of the key. Numbers—what numbers? He tried vainly to send what his grandfather was

telling him, but the best he managed was a group of erratic dots and dashes. He shoved his grandfather from his mind and drifted back to a level of unconsciousness where he was free from pain.

* * * * * * * * * * * * * * * * * *

Julie Meyer's group stopped to rest briefly at a curve in the trail. All of them were perspiring even though the outside temperature was right around freezing. Julie looked at her watch.

"We've gone five miles in two hours! That must be a record on this trail."

Everyone breathlessly agreed with her. All of them could feel their pulses pounding in their temples. It had been an unspoken agreement that they were going to make it to Hunt's Cove as fast as humanly possible. They didn't waste any breath talking— just waited until they felt their hearts slow to a normal rate. Then they shouldered their packs and continued climbing up the steep, snow-covered trail.

* * * * * * * * * * * * * * * * * *

Kim leaned forward eagerly at the table in the communications van. For a minute there, it had almost sounded as if Marc were sending a number, but she couldn't make it out. She added the dots and dashes to the list of scribbles she had been trying to decipher all night.

The roar of a helicopter going overhead filled the van, and Kim stepped outside to watch the National Guard Huey zoom over the treetops.

Chapter 22

"146.52!"

Friday, May 19th
7:00 AM PDST
0700 Military Time

As the giant "Huey" reached the outskirts of the wilderness area, it slowed down and began its methodical grid search. All eyes riveted on the ADF (automatic direction finder) located on the instrument panel. The round dial labeled with North, South, East, West arrows was divided into 360-degree segments.

Carl Blankenship turned the audio up so that Marc's signal could be heard through everyone's headphones. Carl adjusted the tuning on the ADF and headed the helicopter in the direction the arrow was pointing.

"Just how accurate is that thing?" Lieutenant Baxter asked.

"Very accurate," replied Carl. "We've got a buddy who's a real country music fan. He knows the location of every major country music station in the US. So once as an experiment, instead of following usual navigation procedures, he used the ADF to home in on each station and proceeded from state to state that way."

"That's a new one," laughed Lieutenant Baxter. The laughter was welcome because, even though they wouldn't admit it, everyone aboard was understandably tense. They all knew of the approaching weather system. This morning would probably be their last time to search for a couple of days. If the young man were still alive, and they hoped the signal proved that, this was their only chance to find him.

Carl turned the craft slightly to line up with the arrow which was pointing Northeast. Marc's signal had been coming every thirty seconds or so, giving them repeated new bearings.

Now, it became silent. They flew on in a NE direction for five minutes, everyone listening intently for the familiar beep.

"I hope we haven't lost him," Carl said. They were now over the Pamelia Lake area. Carl slowed the helicopter down to just a few knots as they skimmed the snow-topped trees. A group of green- wool-clad Explorers was on a trail below them, and the young people waved a greeting at the helicopter as it climbed and continued to explore a nearby ridge.

"— — — —" Dah, dah, dah, dah—there it was. Another series of inexplicable dashes. Carl wondered briefly if what they were hearing was really being sent by Marc. It certainly didn't make any sense. What if he had left his code key on and a branch caught it, and the wind were sending the code. He had a sudden vivid image of a dead man lying frozen on the ground while gusts of wind bounced a Douglas fir tree branch up and down against a key. He shuddered. It was not an image he would share with the others. They would all know soon enough if the signal were Marc or a tree ghost.

The bearing was directly north now, and the helicopter climbed to make it up over the Crest Trail.

"Come on, Marc, send!" Carl said aloud. He slowed the helicopter, and they hovered above the top of the crest. Everyone peered out the windows, hoping to spot something. The fresh snow was completely trackless. Nothing. Carl nudged the craft on, past the top of the crest.

"•••• — —" Dit, dit, dit, dit, dah, dah. The ADF swung abruptly 180 degrees in response to the signal.

"We've gone past him," Carl shouted. He turned the helicopter around and flew slowly back toward the Pacific Crest Trail.

Back and forth they flew while Lieutenant Baxter marked the approximate area on a grid and section map spread out in front of him. For several frustrating minutes of silence, the helicopter continued in a north-south pattern, waiting for the next signal. When it came, the needle on the ADF held steady to the north and then abruptly retreated 180 degrees as they left the treed area.

"That's it—right back there in those trees," Warrant Officer Blankenship shouted. He turned the craft and made several more passes just to make sure. Luck was with them and each time a beep came through the speaker, it confirmed their bearings.

Lieutenant Baxter marked the information on the map and quickly radioed it back to Sergeant Willis at Base Camp.

They skimmed the treetops, the wind from their rotors blasting snow off in huge chunks. Everyone stared down through the trees, trying to catch sight of Marc on the ground. The trees were so thick that all they could see were the snow-clogged branches.

"Sure wish I could set her down here," Pilot Carl Blankenship said. "But it's impossible. Let's go look for a clearing."

* * * * * * * * * * * * * * * * * * *

Marc groaned. His grandfather was there again. This time he wasn't just insistent, he was angry.

Wake up, Marc! Now! You have to wake up, Marc!

Marc thrashed and turned slightly in his sleeping bag. Pain ran up his body like a speeding truck. He opened his eyes. He couldn't see his grandfather. It was only when his eyes were closed that he could see his image. But he could still hear his voice. Or was it his own? Somehow it seemed like a mixture of his grandfather's bass voice, and his own tone that he used to lecture himself with. Whatever, it was, its message was clear.

146.52—send it now!

Marc grabbed the key and tried to think of "one" in Morse code. His grandfather told him. •— — — — and then he heard the familiar voice sounding it out—*dit dah dah dah dah.*

His fingers. He couldn't move them. Suddenly there was a warm sensation in his right hand. Marc felt the unmistakable feel of his grandfather's leathery hand around his own.

Send, Marc, send!

With agonizing effort, Marc bent the fingers on his hand and clasped the key.

* * * * * * * * * * * * * * * * * * *

Carl Blankenship had just spotted a clearing big enough to land the helicopter about a half mile away from where they believed Marc to be.

What was that?

"Dit dah dah dah dah. Dit dit dit dit dah" Morse code!

He reversed his course and went back over Marc's position. The rest of the crew looked at him questioningly.

"He's sending numbers!" Carl explained. "Real numbers!" He listened intently as the beeping continued. "14 146 14652— why, that's the simplex frequency on two meters! The guy's alive!"

There was a loud cheer in the helicopter as they continued to circle the area. Lieutenant Baxter quickly radioed their happy information back to Base Camp.

* * * * * * * * * * * * * * * * * * *

All up and down the West Coast, "hams" who had been alerted to listen for Marc's signal heard him. Numbers! 14652! Simplex frequency on 2 meters! The lost man was alive!

Almost in unison, they rushed to call the Sheriff's Office number that had been given out on the emergency nets. NX7V, an Amateur Radio operator in Medford, Oregon was the first one to report hearing Marc to the Marion County Sheriff's Department. In the following hour, 143 more "hams" called in with the identical message.

Chapter 23

"TALK TO ME, MARC!"

Friday, May 19th
8:00 AM PDST
0800 Military Time

Sergeant Willis was the first to hear the sound of the hovering helicopter. He came out of the communications van where he had been talking to one of the Posse units and stared over toward Mt. Jefferson.

"They must have spotted something just over the ridge near Hunt's Cove—sounds like they've been in the same place for a long time."

Kim had been over talking to Bill Harris and petting Jake who was standing saddled under the trees. Now, she ran back to Sergeant Willis. Mary came to the door and beckoned them both in. Mary Hammond's normally calm voice sounded excited as she told them the latest news.

"They've found him! They can't see him, but they're sure he's there. He's sent a message in Morse code which Warrant Officer Blankenship understood. The message was '146.52'—'146.52'. Chief Warrant Officer Blankenship says that's the simplex frequency on two meters. He wants to know if anyone has a two-meter transmitter handy. They can't pick that up in the helicopter."

Kim had already pulled hers out of her backpack on the floor. With shaking fingers, she attached the short "rubber ducky" antenna and ran outside to the clearing. Kim pushed the transmit button.

"Marc?" she said tentatively and then louder "Marc!"
Nothing.

Maybe his own call would rouse him, she thought. "Marc! KA7ITR from KA7SJP!" She shouted it several times into the microphone.

"We're not close enough!" she cried. "There's a ridge between us. I've got to get up there," she pleaded, looking at Sergeant Willis.

"Just calm down, Kim. You will." He looked at her flushed face and tearful eyes. He reached out and grabbed her shoulders gently.

"Look, we'll take you up there, but not if you're going to fall apart on us. Are you going to be okay?"

Kim took a deep breath and felt her hammering heart slow a little. She nodded.

"Yes," she said. "I'll be fine—I promise—just get me up there."

"Can you ride a horse?" Sergeant Willis asked.

"Sure," Kim said. "I went to horse camp four summers in a row."

Sergeant Willis motioned Bill Harris over.

"I want your group to take Kim up the mountain. Can you saddle up that extra bay horse you brought?"

"You bet," Bill said. "But let's put Kim on Jake. I'll ride the bay."

It was all done fast enough to suit even Kim. Bill and the other two Posse members conferred with Sergeant Willis about the exact location Lieutenant Baxter had reported. Bill rolled up the map and put it in his saddle bag. Another man, Paul, strapped a metal litter onto his horse, plus an extra sleeping bag. Sergeant Willis nodded his approval. Soon they were on their way.

Bill led out on the bay gelding, and Jake obediently followed. The horses all put their heads down to climb the steep slope. Their noisy breathing made puffs of steam around them in the frosty air.

The trail was rough and more than once Kim thought Jake would stumble. But he always seemed to put his feet down in secure places, and other than a few good bounces, Kim was amazed at how well he handled the trail. At the curve of the first switchback, where they had a direct line of sight to the ridge on the other side of the canyon, they paused while Kim tried to raise Marc.

"Marc!" she shouted several times into the microphone. "KA7ITR from KA7SJP—it's me, Kim!"

There was a groan, soft but clearly audible to the whole group. Bill Harris put his finger to his lips, cautioning the group to be quiet while this extraordinarily important conversation took place.

"Marc! It's Kim! Tell me where you are."

Silence.

"Marc! It's Kim! Listen to the helicopter above you. They can't see you—tell us where you are."

She heard his raspy breathing and then a barely coherent voice started talking. Kim pressed the receiver hard against her ear, trying to make out the words Marc was whispering into the microphone.

"All I got was 'under trees—near creek'," she said, shaking her head. "Does that help?"

Bill Harris immediately radioed the information back to Sergeant Willis at the Base Camp who in turn relayed it to Lieutenant Baxter in the helicopter. In a few minutes, Sergeant Willis's voice came crackling back over Bill's hand-held radio.

"Okay—they can't land there. The first clear spot is about a half mile away. They're running out of fuel so they need to set down or return. I want them to land so they'll be available for transport.

"They say there is a steep talus slope just past Hunt's Cove. There's a creek at the base of it that follows the line of the trees. They think he's under those trees somewhere. I'm going to give you exact grid and section markings, but it should help to try following the creek."

They waited while Bill marked down the bearings that Sergeant Willis gave him, and then they continued up the trail, urging their horses to go as fast as they could. Kim kept her two-meter rig turned on. Bill Harris turned around and offered to take her reins while she concentrated on transmitting. Kim gratefully accepted.

Pressing the radio to her ear, she listened and then brought it back down to her mouth to transmit.

"Marc—it's Kim! We're coming to get you. Hold on!"

"Kim?" Marc's barely audible voice replied. "Kim?"

"Good girl," said Bill Harris. "Keep talking to him—keep him awake."

The sound of Marc's labored breathing stopped, and Kim guessed he had switched his rig back to receive by letting up on the transmit button.

"Keep talking to me, Marc!" she begged. "We're coming to get you. You'll be okay. Can you tell us where you are?"

Silence.

"Marc!" Kim heard herself screaming into the microphone. "Don't go to sleep, Marc! Talk to me!"

* * * * * * * * * * * * * * * * * * * *

It was too much. The effort of talking had exhausted his last reserve of energy. The two-meter rig fell from Marc's hand onto the ground as he sank back into a white void that swirled and tumbled and became dark.

* * * * * * * * * * * * * * * * * * * *

Kim was openly crying as they urged their horses to quicken their pace up the mountainside. She had shouted into the microphone until she was hoarse. No answer. She was sure they were too late. Marc's life had slipped away.

The rest of the group didn't say anything. Just the noisy breathing of the horses and creaking saddles broke the sunlit silence of the morning.

Bill stopped the group under some trees. Mt. Jefferson was directly across from them. Its snowy magnificence looked almost blue-white in the bright sun. The horses, grateful for the brief stop, nosed each other and shifted their weight from foot to foot.

"Why are we stopping?" Kim demanded, almost hysterical.

"Time to check in," Bill Harris explained. Everyone in the group seemed to be sympathetic to her emotion. Another rider reached out and touched her arm gently. Kim took a deep

166

breath, ashamed of her outburst. She had promised Sergeant Willis she wouldn't do this.

Mary Hammond was talking to Bill, telling him that the Explorers had reached Hunt's Cove and were proceeding along the creekbed toward where the helicopter thought Marc was.

"I think we can catch up with them," Bill said as he signed off.

Once again, they spurred their horses into action. Kim kept trying to transmit and listen to Marc, but there was no response. She looped the cord attached to her two-meter rig around the saddlehorn and told Bill she could take the reins back.

"I'm okay, really I am," she told him. "We can probably go faster if I guide my own horse."

He didn't argue with her and urged the bay forward. Soon, the trail leveled out and they passed a couple of beautiful snow-filled meadows. At the second one, Bill pointed briefly and said, "That's Hunt's."

He led the group down an incline through the meadow, and they splashed through the icy stream that bordered it.

"I guess we're supposed to follow this," he said, pointing to the stream. "I bet Marc tried to climb that!" He pointed to the snow-covered steep slope beside them. A few jagged rocks stuck out through the snow.

"The Crest Trail is up on top there. Don't know where he was heading, but perhaps he was taking a shortcut."

They kept the horses along the stream. Dense forest began about twenty yards from the water, and a couple of the riders rode through the trees, shouting "Marc!" as they went.

Suddenly, there was a triumphant yell far ahead through the woods.

"That's the Explorers!" Bill said, excitedly. "They've found him!"

The terrain was rough, but he forced his horse into a semitrot. Kim felt nervousness and fear washing over her. What had they found? Was he still alive?

A loud voice coming through her two-meter rig almost made her jump out of the saddle.

"Horse Posse #1—this is Explorer Team #2. Can you hear me?"

"Yes—go ahead," Kim pleaded.

"We've found Marc and he is alive—I repeat—he is alive!"

The group rounded a bend of the stream and spotted the Explorers under a particularly dense thicket of tall fir trees. One of them was talking on Marc's two-meter rig. He waved when he saw the approaching Posse and turned back to help the others.

Bill and one other Posse member were already off their horses and running toward the scene. The third member got off to stay with the horses. Kim, suddenly feeling awkward, dismounted, ran to the group, and then stopped. She couldn't even see Marc —there were so many people around him.

She heard some low groans and then as a girl about her age moved over, she had a glimpse of Marc. The first thought that struck her was that he looked familiar—like she had known him all of her life. His pale, frostbitten face was a ghostly resemblance to the photo Mike had shown her. Anxiously, she watched the first aid efforts under way.

Two Explorers had actually climbed into the sleeping bag beside him, and a third was cupping his hands over the frozen portions of Marc's face. Someone had spread an extra sleeping bag over them.

Kim turned as she heard footsteps crunching through the snow. A tall blond man in a National Guard uniform and a shorter gray-haired man wearing a sheriff's officer uniform were approaching.

Bill Harris greeted them. It was Lieutenant Baxter and the crew chief from the helicopter.

"We've set down about a half mile from here," Lieutenant Baxter said. "How's he doing?"

As if in answer, there was a loud groan from Marc. One of the Explorers had pulled back the side of his sleeping bag and was attaching a splint to his injured leg. They were working as quickly and gently as they could, but the excruciating pain of anyone touching his swollen leg was breaking through Marc's unconsciousness.

"Sorry, Marc," the young man working on his leg said to him. "I'm just about done."

Marc opened his eyes fully and stared up at them. Kim saw a place on the ground near his head, and quickly knelt beside him. A girl on the other side of him had one of his chalky-white hands inside her jacket next to her body.

"Take his other hand," she told Kim. "Put it inside your jacket under your armpit."

Kim did as she was told and felt the icy flesh of his hand through her flannel shirt. Marc turned his head slightly and looked up at her. His eyes looked glazed.

"It's Kim," she told him. "You're going to be okay, Marc."

"Kim?" His lips formed the word, but no sound came out.

"He's beginning to rouse," Bill Harris told Lieutenant Baxter. "I think we can stabilize him here and then transport him."

One of the other riders had already gathered some firewood and started a fire just a few feet from them. The heat of the flames felt good to Kim as she sat there absorbing some of the cold from Marc's body.

Gradually, he became more and more awake. It was obvious that he was in terrible pain. Sweat broke out on his forehead as he grimaced.

Lieutenant Baxter held a warm cup of tea to his lips.

"Drink this, son," he said, slipping a hand behind Marc's neck and raising his head a few inches.

Marc opened his parched-looking lips slightly and tried to take a sip. Most of it ran down the side of his chin. Lieutenant Baxter gently blotted the tea away with a handkerchief and waited a few minutes and then tried again. This time, Marc swallowed several mouthfuls.

He started talking more, but the effort exhausted him, and he just lay quietly, watching the activity around him. Occasionally, he glanced up at Kim, and she smiled back at him each time.

Satisfied that his body temperature was rising, the group began readying him for the transport back to the helicopter. A metal stretcher was laid next to Marc. The splint on his leg was padded some more to make his leg as immobile as possible.

"Ready?" Julie Meyers asked.

"Ready," all four Explorers replied. They had positioned themselves on both sides of Marc, and gently they picked him up, sleeping bag and all, and transferred him to the stretcher. Kim saw him clench his jaws as they moved him and a slight moan escaped his lips.

She bent over and touched his shoulder.

"I'll see you later," she said softly.

"Okay," Marc answered. Then his eyes opened to full alertness. "Kim, would you get my rig?—it's a good one. I'd hate to lose it."

"Done," she promised.

The Explorers picked up the litter and started back along the stream. Two members of the Posse, Lieutenant Baxter, and the helicopter Crew Chief followed.

"They'll trade off carrying him," Bill Harris explained to her. "A person gets pretty heavy when you have to carry him in terrain like this."

Kim watched as they disappeared through the trees. She had overheard Lieutenant Baxter say that he would be flown to the National Guard Airfield and then transported by ambulance from there. She turned to Bill and smiled, really smiled for the first time in days.

"I know just how you're feeling," he said to her. "I feel that same thrill every time we find someone alive. And to tell you the truth, I had my doubts about whether we were going to find this guy alive."

A glob of snow, melting in the bright sunlight, fell and hit Kim squarely on the head. She laughed and looked up at the trees.

"Well, look at that!" she said pointing to Marc's antenna strung over a branch. She bent over and picked up his radio gear lying on the ground.

170

A flashlight with the end smashed out of it was attached with two wires to his HW-9.

"That's a strange power supply," she commented. "Really strange!"

Out of curiosity, she opened his backpack and rummaged around. She pulled out a 12 volt battery.

"Wonder what was wrong with this one?" she said. She felt around some more in the jumble of stuff and pulled out the crushed headphones.

"Just as I thought. Wow. This is amazing!"

Bill watched with interest as she examined the radio gear and then stuffed it all back into Marc's backpack. They tied it on the back of Jake's saddle, put out the fire, and then mounted for the ride back to camp.

Chapter 24

"MEET ME SUNDAY?"

Saturday, May 20th
2:00 PM PDST
900 Zulu

Kim ran her fingers through her brown curly hair
self-consciously as she paused in the doorway of
Marc's hospital room and peeked in. She was relieved
to see that he had been moved out of Intensive Care. His face
had regained a normal color except for some white patches of
frostbite. He appeared to be sleeping peacefully.

His broken leg was swathed in bandages and suspended
in traction. It would be a couple of weeks before it could be set,
the doctor had told them last night.

After the dramatic rescue yesterday, it had been noon by
the time Kim got back to the Base Camp on horseback. Mike
had arrived and was waiting anxiously for their return. In an
excited jumble of words, Kim had told him about the rescue
and had reassured him about Marc's condition.

"He was really pale and sort of half-asleep, but he tried
to talk to us. His leg looked pretty bad, but everyone who was
giving him first aid seemed to think he would be okay. They
said his hands and feet and face were frostbitten but not as
bad as they could have been. He's alive! That's the important
thing—he's alive!"

Mike had agreed wholeheartedly and had gripped Kim's
arm and waltzed her around in a circle.

"Well, what are we waiting for?" he said happily. "Let's
get down to the hospital!"

Mary Hammond, Sergeant Willis, and Sharon Hansen
had watched the two of them fondly as they grabbed their stuff
to leave. Sharon had just reported in for duty, and now it

seemed like her only duty would be to take her 80-meter transceiver home.

"Thank you—thank you everyone!" Kim yelled. She ran over and gave Bernie, who was cooking some stew for the hungry Explorers, a big hug. Next, she went over to tell Mary and Sharon goodbye.

"You were great, both of you," she said sincerely. Mary was helping Sharon take down the 80-meter antenna wire and was asking her questions about Amateur Radio.

Sharon laughed. "Here, I've been working on Mary for years to become a ham, and then you come along, Kim, and suddenly, she's a convert! We ought to get you to speak at ham conventions!"

Kim grinned and gave them both bear hugs.

Sergeant Willis was helping to unsaddle Jake and load him back into the trailer.

Kim came up behind him and waited for him to turn around and notice her. He smiled at her and stuck out his hand to shake hers, but Kim wasn't in any handshaking mood. She gave him a big hug and kissed him on the cheek.

"Thank you, Sergeant Willis for everything. Thank you for leading such a good search and thank you for letting me help."

"Help, my eye, young lady. Kim, you saved the day—you and that Amateur Radio of yours!" He tousled her hair fondly.

Kim grinned. "And who flew the helicopter? And who led the searchers? And who organized the whole thing?" she asked.

"Okay," he said smiling. "You're wonderful, the Explorers are wonderful, the Posse is wonderful, the National Guard is wonderful—satisfied?"

"And you're wonderful, too," Kim added.

"Enough!" Sergeant Willis said. "Don't you two need to get to the hospital to see Marc?"

They waved and ran to Mike's car. Kim thought the valley had never looked more beautiful as they came down out of the snowy mountains and saw the glistening rain-soaked hay and wheat fields that formed the flatlands of the Willamette Valley.

They drove into the hospital parking lot and ran to the Emergency Room entrance. A nurse told them that Marc had been stabilized and transferred to Intensive Care. They took the elevator to the third floor and walked to the nurses' station outside the Intensive Care units.

A dark-haired doctor who was busy writing on a chart looked up when he heard the two of them ask about "Marc Lawrence."

"Are you the girl who was in on the rescue?" he asked.

Kim nodded. The doctor introduced himself as Dr. Malbey and led them to a nearby lounge.

"Your friend was really lucky," he said. "Even so, he's very ill. I think if he hadn't been rescued today, it would have been too late."

Then he explained more about Marc's injuries. He had a bad break in his leg which had become compounded. They weren't sure if the bones had separated at the time of the injury or later. At any rate, it was going to take traction to align them before the leg could be surgically set.

"And he has a serious infection. We're not sure if it's from the leg or all the cuts and scrapes on his body. We've started him on antibiotics, and he should respond. His hands and feet are frostbitten, but at this point, it doesn't look like he'll lose any fingers or toes."

Kim and Mike nodded at all this information. The important thing was that Marc was alive. They clung to the doctor's final words.

"Barring any complications, he should make a full recovery."

"Can we see him?" Kim asked.

"Well, you can go to the doorway and peek. He's asleep—he was pretty exhausted after all that transporting and in quite a bit of pain. We've given him some pain medication. I imagine he'll be able to visit with you some, tomorrow."

174

Quietly, they had gone to the curtained area and stuck their heads in to look. Marc still looked pale but better than he had on the mountain. An I.V. dripped into one arm, and his leg was suspended in traction. He was groaning slightly, his head turning from side to side on the pillow.

The Intensive Care nurse tiptoed over to talk to them briefly.

"He's all right—really. His temperature has come back up to normal. In fact, he's running a bit of a fever, but the antibiotics should take care of that. Why don't you come back tomorrow morning?"

Kim and Mike had driven home. A sudden feeling of joyful exhaustion had settled on both of them.

"I feel like sleeping for 12 hours myself," Kim admitted to Mike in her driveway.

"Why don't you?" Mike urged. "Listen, what time are you going to go to the hospital tomorrow? I'll come back up."

"Oh, I guess about noon," Kim said. "Why don't you just meet me there—that's closer for you anyway."

Mike agreed and they said goodbye. Kim went inside. Her parents were both at work. She wrote a long note for them on the kitchen table, took a hot shower, and collapsed into bed.

Around 8:00 PM, she woke up and called the hospital. Marc was doing just fine. He had awakened briefly and sipped some hot soup. Kim breathed a sigh of relief and went downstairs to visit with her parents and brother and eat dinner. She was surprised that she was still tired, and before long, she went back to bed again.

The sunlight streaming through her window woke her at daybreak. She got up, put on makeup and combed her hair for the first time in days, and went downstairs.

Her mother was busy cooking pancakes. Kim sat down at the kitchen phone and called the hospital.

"Marc is doing fine," the Intensive Care nurse told her. "He's awake and has had some breakfast. We're just getting ready to move him to a regular room."

"Great," said Kim happily. "When are visiting hours so I can see him?"

"Anytime after eleven. We ask that you stay for just a few minutes though. He's a pretty tired young man."

It had been all Kim could do to wait until eleven. Now that she was here, though, she felt suddenly shy. Marc was asleep. One arm was hooked up to the I.V., but the other one lay limply on top of the bedspread. Quietly, Kim tiptoed in and sat down beside him. She reached out and put her own hand gently in his.

His eyes fluttered open and focused on her face.

"Kim?"

"Yes, Marc. It's Kim." They both seemed at a loss for words. They looked at each other and finally both smiled at the same time.

"You're even prettier than I imagined," Marc said.

"Oh," said Kim, embarrassed, "Besides being the toughest guy around, plus about the most ingenious, you're full of flattery, too."

"Not flattery," Marc said softly, "just truth."

Kim let go of his hand and rummaged in her purse.

"I've got something for you," she said. She pulled his two-meter rig from her purse and handed it to him.

"How on earth were you able to do everything you did?" she asked.

"I'm not really sure of that myself," Marc said. "I think I had some help."

Kim looked at him questioningly.

"I'll tell you about it some time," he promised.

The nurse came into the room.

"Marc, your parents just called from Austria. They got hold of your roommate this morning and heard about what happened. They send you their love and said they'll be returning tonight."

Marc smiled at the information.

"Man, my mom's going to come unglued when she hears the details of this escapade."

The nurse laughed but then said reassuringly, "I think she'll just be glad that you're okay."

She looked at Kim and told her not to stay too much longer. Kim could see that Marc was looking tired, but when she suggested leaving, he protested.

"When am I going to see you again?"

"I'll be back, I promise. Mike's coming pretty soon. Tell him I've already been here. I'll call him later."

"You didn't answer my question, Kim. When will I see you?"

"How about tonight?" she said.

"It's a date," Marc said, smiling. "Say," he said as she got up to leave. "Do you have two meters in your car?"

"Yeah," Kim said. "Why?"

"Monitor for me."

"You're supposed to sleep," Kim lectured him, trying to look stern, but laughing instead.

"Just monitor for me—okay?"

Kim gave his hand a squeeze and walked down the hallway and outside to the parking lot.

She got in the car and turned on her two-meter rig. She was barely out of the parking lot when the call came.

"KA7SJP from KA7ITR."

Happily, Kim grabbed her microphone to answer.

Author's Note

I was four years old when my father took me out to the ham shack to talk to Santa one chilly December morning. I remember standing transfixed as a deep booming "Ho ho ho" came across the airwaves. My belief in two wonders—Santa Claus and Amateur Radio—was confirmed that morning.

At times, Amateur Radio may seem as magical as Santa, but there is nothing make believe about Amateur Radio. It is a very real, fascinating hobby, attainable by anyone who has the desire to communicate with the world.

Because of Amateur Radio, I have talked with loved ones thousands of miles away. Because of Amateur Radio, I have met countless new friends. Because of Amateur Radio, I am ready to provide emergency communications in case of a disaster.

And because I hope you will want to become an Amateur Radio operator too, I wrote a novel. But I didn't do it alone. I had technical advice from the following people, and I am extremely grateful for their help and support.

Sergeant Jerry Blaylock, Marion County Sheriff's
    Department
Sidney Dawson
Billie Delaney, WB7RHO, Jeep Patrol
Mildred Flickinger
Robert Goetz, M.D.
Chief Warrant Officer Ken Hiigel, N7BAX, Oregon
    National Guard
Lenore Jensen, W6NAZ
Stephen Jensen, W6RHM
Mary Jo Lundsten, Trail Guide
Chief Warrant Officer Bob McGowan, KD7GP,
    Oregon National Guard
Fred Molesworth, AF7S
Hollie Molesworth, KA7SJP
Richard Ries, Explorer Post 18

John Rigby, Expert Backpacker
Captain Leon Riggs, Marion County Sheriff's Department
Don Sanders, Oregon Mounted Posse
Bill Toman, E.M.T. Instructor
Bob Wall
David Wall
Michael Wall, KA7ITR

Hope to hear you on the air!

73,

Cynthia Wall, KA7ITT